Beyond Plastic

Trends in the Payment Card Industry

Michael A. Brooks

authorHOUSE®

AuthorHouse™
1663 Liberty Drive
Bloomington, IN 47403
www.authorhouse.com
Phone: 1-800-839-8640

First published by AuthorHouse 1/18/2010

ISBN: 978-1-4490-7243-8 (sc)
ISBN: 978-1-4490-7244-5 (e)

Library of Congress Control Number: 2010900255

Printed in the United States of America
Bloomington, Indiana

This book is printed on acid-free paper.

Contents

Introduction

The crumbling of the subprime mortgage loans and subsequent failures of banks and investment firms sent shockwaves that reverberated throughout the nation and the world in 2008. Fannie Mae and Freddie Mac failed so miserably that they were taken over by the U.S. government. Washington Mutual, Wachovia Bank, Bear Stearns, and Lehman Brothers were among the many well-known financial companies that survived only through acquisition. As a result of this economic debacle, ten of the largest United States banks were — and still are — required to set aside $75 billion to prepare for potential losses that could result from a deepening recession. Then along came the government's helping hand to AIG, further weakening confidence in the U.S. economy. The auto industry stood in line waiting for its handout, too, and now a part of General Motors is owned by U.S. taxpayers.

To make matters even worse, many investors saw their portfolios drop about 40% over several months. In early 2009, investor confidence remained shaken. By Q3, the market had started to rebound, though the upward trend was anything but steep.

Bernard Madoff, a former Wall Street icon, had managed to bilk investors out of $50 billion. As a result of his Ponzi scheme, wealthy investors lost their fortunes, charities watched their capital shrink, and middle class workers' retirement portfolios evaporated practically overnight. Add to this scenario the bad news of declining home values, a rising foreclosure rate, soaring unemployment, and plummeting interest rates, and you have a financial meltdown not seen since the Great Depression.

What lies ahead? Growth will likely be sluggish: The economy typically expands at an annual rate of 3.5%, yet the United States appears more likely to grow at only 2–2.5% over the next few years. [1]

---

1   David Lynch, "U.S. Must Face Years of Sluggish Growth," *USA Today*, May 8, 2009.

This slowing economy impacts the credit business, and the credit card industry could face write-offs of $60 billion by the end of 2009.[2] This bleak economic picture does not mean, however, that payment card industry leaders will fall to their knees, groveling for a handout, like some of the major automobile companies, investment firms, or banks have done. It doesn't mean that consumers will suffer in the long term by having to give up credit cards, debit cards, and other convenient forms of payment.

What it does mean is that the payment card business — which, for the purposes of this book, refers to the industry that provides credit cards, debit cards, and prepaid cards, such as gift cards — will need to look at new markets, emerging technologies, and innovative programs to attract and retain the right customers. While many other business sectors will see their fortunes fade, the leading companies in this industry are expected to be profitable, with the annual revenue from the credit card business anticipated to be about $14 trillion. This revenue will ultimately play a more significant role in contributing to the long-term economic growth of the United States. Even the newly passed *Credit Cardholders' Bill of Rights Act,* which is discussed in Chapter 3, will not stop the growth of the credit card business (For more information about this bill, visit www.govtrack.us/congress/bill.xpd?bill=h111-627).

In 2008, consumers began spending less as they tried to make their way out of ever-increasing debt. Some payment card companies, however, have prospered during this time, and will continue to do well, even while many other businesses are struggling. Why? Because, in addition to improved business *efficiencies*, which helped Visa and other companies become profitable, business *opportunities* will not go away. People will continue to spend money and to borrow, although they probably won't do so at the same rate as in the past.

Credit and debit cards are the easiest, most cost-effective and expedient means of loaning money to people and businesses, especially during

---

2   Eric Dash, "The Last Temptation of Plastic," *New York Times*, December 7, 2008.

tough economic times, when other loan options are not available. Credit will be more difficult for consumers to obtain, however, as banks crack down on lending, particularly in light of new consumer safeguards related to credit card legislation. People want convenience, and the payment card industry will provide this service. But consumers will change their approach to spending. They will, over time, be forced to become more responsible and, ultimately, more conservative in the way they look at credit, spending, and their approach to saving. In fact, savings rates are already beginning to increase for the first time in years.

Despite the economic downturn in 2008 and its projected continuation throughout 2009 — in which Oppenheimer & Company analyst Meredith Whitney predicts that issuers will cut more than $2 trillion in credit limits by the end of the year[3] — opportunities for growth still exist in the payment card industry. This growth will occur with an increase in credit transaction volume, the purchase of prepaid cards, debit transactions, and other business. These activities, combined with other types of electronic processing, will generate transaction-based business to help increase consumer and business spending in the United States.

The amount of business that takes place on the Internet continues to expand, and that will help fuel the growth in credit card transactions. CNN Money reported that e-commerce would total about $259B in 2007 alone.[4] Transaction-based purchasing is the lifeblood of Internet-based businesses and will only continue to escalate as more companies increase their presence on the web.

The payment card industry will be a shining light in a struggling economy. Industry leaders — such as Visa, PayPal, and others — will offer a reliable source of significant revenue that is not dependent upon oil or other natural resources. For example, in April 2009, a time when many businesses suffered major setbacks, Visa's second quarter earnings beat estimates.

---

3   Ellen Cannon, "The New Credit Card Future," www.bankrate.com, March 12, 2009.
4   Ed Kim, "Visa, MasterCard Risk Ramped Up Competition," www. seekingalpha.com, March 27, 2008.

PayPal Merchant Services is a major source of growth for PayPal, which is on track to reach around $3.3 billion in 2009, with an annual growth of 35% — all at a time when many companies' stocks have been completely battered.[5] PayPal will play an even stronger role in this new economy, which is increasingly moving toward electronic payments and away from cash. According to industry analyst Ed Kim, "PayPal is accepted in other e-commerce sites, making it a major contender to Visa and MasterCard for payment services on the Internet."[6]

The expansion of electronic payments will ultimately contribute to the greening of our economy as we print and ship fewer paper checks. Electronic processing will save trees and reduce the environmental toll of paper mills and mail trucks, while simultaneously reducing administrative costs. These expenses are significant, too: The cost to print and store a multi-page statement can be more than $1 each.[7] Multiply that by the 1.5 billion U.S. credit card account statements mailed 12 times a year, and you save over $16.8 billion a year in printing costs alone.

What about the consumers? How can they possibly benefit from increased credit tightening at a time when unemployment is climbing upward? How can they become more responsible when the American culture has always been one of not just "keeping up with the Jones'," but also surpassing them with the hottest car, biggest house, and the most fashionable designer clothes? How can consumers possibly benefit when the children of Baby Boomers will be the first generation in a long time that is likely to be less successful financially than their parents? How can life get better for consumers at a time when a growing percentage of adult children have to move back home with their parents because they can't afford a place to live?

---

5   "PayPal Growth Counterbalances eBay's Slower Growth," www. seekingalpha.com, July 24, 2009.
6   Ed Kim, "Visa, MasterCard Risk Ramped Up Competition," www. seekingalpha.com, March 27, 2008.
7   Kathi Plymouth and Jodi Martin, "Bill Payment Trends: Major Shifts in Consumer Behavior Require Comprehensive Planning," www.firstdata.com, 2009.

What will happen to small businesses over the next few years? They help perpetuate the American Dream, and yet small business owners will face new challenges due to their high reliance upon credit for financing. Will they be able to find the capital needed to foster growth, or will their businesses struggle in an attempt to merely survive?

What new technologies on the horizon will have the most dramatic impact on consumer spending? How will the payment card industry innovate to help increase the number of transactions that can be completed in a given time while simultaneously improving the security of those transactions? How will the industry make it even easier for people to make their purchases from anywhere at any time? What personal information will consumers have to give up just for the privilege of transacting business?

This book will examine how the current economic climate will cause cultural shifts in the way people and businesses spend money and use credit. It will look at how this change impacts the growth of debit cards and other forms of electronic payments and technology. This book will explore how innovations in technology, payment programs, and new markets — such as those for mobile technology and related innovations — will lead the way in the new economy.

Ultimately, in the long term, consumers will be better off, because they will become more aware of how to use credit wisely. Banks will continue to be more cautious in extending credit, while leading payment card companies — such as Visa — will reach expanding markets in India, China, Latin America, Eastern Europe, Malaysia, and other countries. The expansion of payment cards to other nations will help to make their markets part of the new economy — all while bringing significant revenue back to the United States.

This book will also explore the impact of credit card and payment card regulations and fees, and explain how to prevent fraud and identify theft — at a time when electronic transactions are growing dramatically year after year. It will discuss the strategies of companies that will compete for business in this industry, and identify which of those are most likely to survive and prosper. In addition, the book will address the changing face of money and what this means — the growth of mobile transactions, biometrics, payment cards, smart cards, contactless devices, and more.

The news headlines of today tend to focus on consumers who have been victimized by high credit card rates and the negative impact of credit card debt. In fact, the average outstanding credit card debt for households that have a credit card was nearly $10,700 by the end of 2008.[8] Yet, much less attention is given to the benefits of the payment industry: Without credit cards, many people would not be able to start a business, pay their medical bills, pay college tuition, cover essential expenses during times of unemployment, or have money available for other emergencies. In fact, 59% of small businesses reported relying on credit cards to provide business capital, and credit cards are often considered the ideal platform for small business loans.[9]

While the easy use of credit has contributed to consumer overspending in the past, it has also given people the opportunity to improve the quality of their lives by helping them pay for vacations, entertainment, education, necessities, and other experiences or conveniences. By putting expenses on a credit or debit card, rather than using cash, consumers can have a more effective way of tracking their spending. Having a good budgeting and tracking system is an important step toward responsible money management.

Credit is vital to our economy, and we don't need to think of it as a dirty word. While credit, in the recent past, has all too often been used by consumers to pay for what they *want*, our changing economy will facilitate a cultural shift. People will begin to use credit more for what they *need*. The instant gratification associated with credit will be less important to consumers as they learn how to take a more responsible approach to managing their finances. Though the credit industry will be more tightly

8   Ben Woolsey and Matt Schulz, Credit Card Statistics, Industry Facts, Debt Statistics www.creditcards.com, September 10, 2009.
9   *NSBA Small Business Credit Card Survey*, National Small Business Association, www.nsba.biz/docs/09CCSurvey.pdf, 2009.

regulated going forward, these efforts will benefit consumers by controlling interest rate increases and making customers more aware of the terms of their contracts.

While many companies will try to capture the payment industry business, the most innovative companies — such as Visa, MasterCard, and PayPal — will continue to prosper, even in challenging economic times. They offer a service that is both in demand and helping to fuel growth. They have the platforms and the infrastructure to make the purchasing experience more secure, faster, and more effective, both for the consumer and the business. The payment card industry may face some hurdles, but it will remain strong. It will help the United States on its road to economic recovery and provide the services that consumers need and want.

Chapter 1 - An Overview of the Credit Industry

Reviewing a brief history of the credit card industry is essential in order to understand how it will evolve over the next decade and to comprehend its impact on the economy and consumers. This book will review how the credit industry grew into the powerhouse it is today (see Figure 1) and will introduce some of the key companies that got it here. This chapter explores how the credit industry works, how money moves from one institution to another, and who collects fees from whom.

2009 — a	a.) Credit Cardholders' Bill of Rights passes
2008 — b	b.) Visa completes the largest initial public offering in U.S. history and begins trading publicly on the New York Stock Exchange under the ticker symbol "V." Discover Financial Services reaches $2.75 billion settlement agreement in antitrust dispute with Visa and MasterCard.
2007 — c d	
2006 — e	c.) Visa announces the completion of the company's corporate restructuring, creating a new global corporation called Visa Inc.
2004 — f	d.) Visa launches the Visa mobile platform, a business and technology framework for facilitating the adoption of mobile payments and value-added services.
2002 — g	e.) **MasterCard transitions to a new corporate governance and ownership structure. MasterCard Inc. begins trading on the New York Stock Exchange under ticker symbol MA. Google Checkout is announced, a way to pay for online purchases.**
2001 — h	
2000 — i	f.) Visa's global debit volume surpasses credit. Diner's Club and MasterCard finalize groundbreaking alliance.
1999 — j	g.) **MasterCard merges with Europay International to create MasterCard International. eBay buys PayPal for nearly $2 billion.**
1998 — k	h.) Visa's annual global sales volume reaches the $2 trillion milestone.
1997 — l	i.) American Express celebrates its 150th anniversary.
	j.) Visa conducts the first euro transaction using a payment card.
1995 — m	k.) Peter Thiel and Max Levchin founded a company that later becomes PayPal. First Data provides transaction processing services worldwide.
	l.) **Visa's annual global sales volume reaches US $1 trillion. Master-Card launches the Priceless®, award-winning advertising campaign.**
1993 — n	
1992 — o	m.) Visa co-develops industry-wide chip card specifications.
	n.) Visa applies state-of-the-art technologies to payments, potentially reducing the incidence of card fraud by giving Visa member banks smarter and timelier data about suspicious transactions. Visa offers the first international prepaid card.
	o.) **First Data spins off from American Express and goes public.**
1989 — p	p.) **Life Magazine names Frank McNamara, who invented the first credit card, one of the 100 most influential Americans of the 20th century.**
1987 — q	q.) **MasterCard is the first-ever payment card issued in the People's Republic of China.**
1986 — r	r.) Visa becomes the first payment card system to offer multiple-currency clearing and settlement, increasing transaction efficiency. American Express' earnings exceed $1 billion for the first time. Discover card is unveiled nationally during the 1986 Super Bowl.

Year		Event
1985	s	s.) MasterCard acquires an interest in Europay International's predecessor, EuroCard.
1984	t	t.) **Diner's Club creates the first rewards program.**
1983	u	u.) **MasterCard introduces hologram security device, an industry first. Visa launches a global ATM network, providing 24-hour cash access to cardholders around the world**
		v.) **Diner's Club is the first travel and entertainment card used in China.**
1980	v	
1979	w	w.) **ICA is renamed "MasterCard."**
1976	x	x.) BankAmericard changes its name to Visa.
1975	y	y.) Diner's Club introduces the first corporate card program in the industry.
1974	z	z.) The International Bankcard Company (IBANCO) is formed to administer the BankAmericard program internationally.
1973	aa	aa.) NBI launches the first electronic authorization system, followed a year later by an electronic clearing and settlement system, the precursor to VisaNet.
1971	bb	bb.) First Data Resources (FDR) is incorporated to become a for-profit organization providing processing services to the Mid-America Bankcard Association.
1970	cc	
1969	dd	cc.) Visa is incorporated in the state of Delaware in 1970 as National BankAmericard Inc. (NBI).
1968	ee	dd.) **ICA acquires exclusive rights to the "Master Charge" name and the trademarked interlocking circles. Diner's Club is the first charge card in Russia.**
1966	ff	
1961	gg	ee.) Diner's Club has 1.8 million card members.
1959	hh	ff.) The Interbank Card Association (ICA) is founded.
		gg.) Diner's Club replaces its paper card with plastic.
1958	ii	hh.) Diner's Club becomes the first charge card to offer confirmed hotel reservations.
		ii.) Bank of America launches the BankAmericard with an innovative "revolving credit" feature. American Express issued its first charge card.
1951	jj	
1950	kk	jj.) 20,000 people become Diner's Club members.
		kk.) Diner's Club issues the world's first charge card and becomes an alternative to cash.
		ll.) American Express launches the money order business.
1882	ll	mm.) American Express is established as a delivery business.
1850	mm	

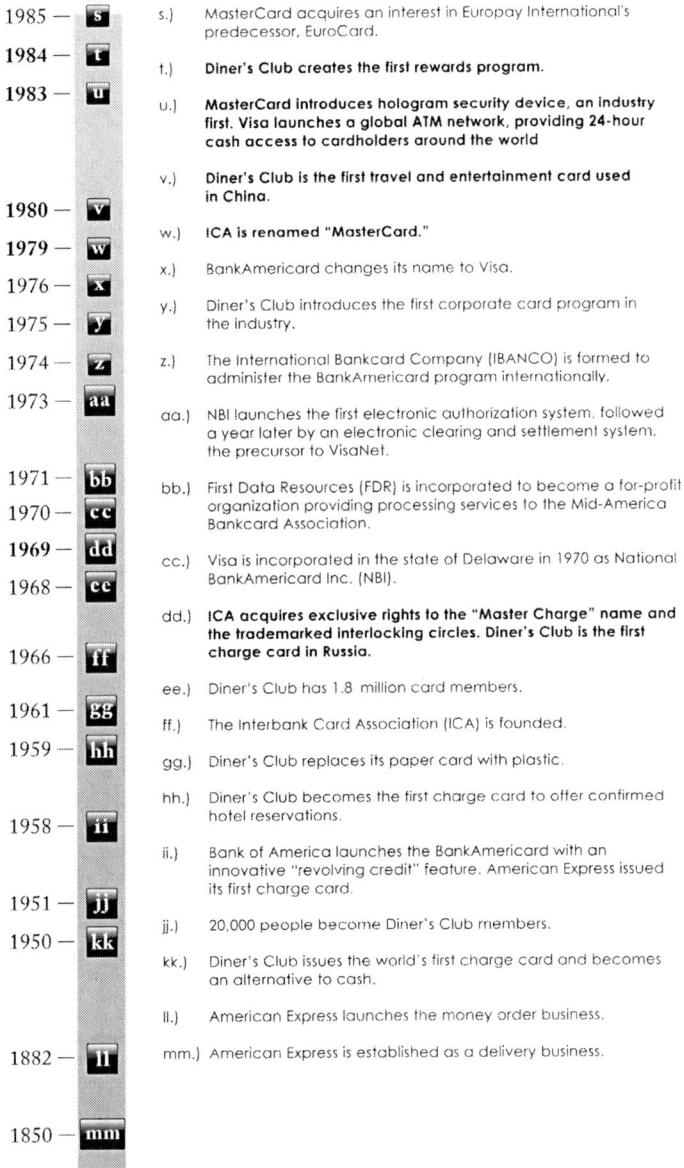

Figure 1: Significant Events in the History of the Credit Industry.
(*Sources: Visa.com, AmericanExpress.com, DinersCard.com, eBay.com, MasterCard.com, Google.com, and FirstData.com*)

A Brief History of Modern Credit

Modern credit first began taking shape around the 1920s, when stores saw a way to ensure customer loyalty by providing the option of paying for purchases over time. The actual credit card made its debut in 1946, when Brooklyn's Flatbush National Bank issued a credit card to select customers. Just four years later, Diners Club also issued a credit card for customers to use at restaurants. At that time, owning a credit card was considered to be a prestigious privilege for a select group of people who used the card to pay for business expenses or for dining at designated restaurants.

In 1958, American Express issued its initial charge card, which became the first widely accepted credit card. However, the company focused primarily on business customers and travelers' checks. At the start of the 1960s, only 7% of American households had a credit card. Five years later, more than 1 million American Express cards were in circulation, with some 85,000 merchants accepting them.

MasterCard, created by a group of banks,[10] including United California Bank, Wells Fargo, Crocker Bank, and Bank of California was first formed as the Interbank Card Association in 1966.[11] While banks competed with each other for merchants and consumers, they worked together by setting operational standards for credit cards.

In 1958, Bank of America introduced the BankAmericard — the first card that let people pay their balances over time.

By 1968, Interbank had 286 banks. Interbank charged a modest entrance fee and a small annual fee.[12]

---

10 MasterCard Credit Card Company, www.newsdial.com/money/credit-cards/mastercard.html
11 www.MasterCard.com.
12 David S Evans and Richard Schmaensee, *Paying With Plastic: The Digital Revolution in Buying and Borrowing*, MIT Press, January 2005.

In 1969, Interbank Card Association changed its name and became known as Master Charge. The company made its final transition to become MasterCard in 1979.

By 1970, Bank of America had connected its system to National BankAmericard, Inc.

In 1977, Bank of America changed its card name to Visa[13] and, eventually, a new leader in the credit card industry was born.

The Visionary Behind Visa

Visa was the vision of Dee Hock, a remarkable businessman who gained recognition by paving the way for the rise of the credit card industry in the United States. Hock worked for the National Bank of Commerce when the bank became the licensee of Bank of America's BankAmericard. He left Visa in 1984 and was inducted into the Business Hall of Fame in 1991. He also received recognition from *Money* magazine in 1992, when he was listed as one of eight individuals who had most significantly shaped the way people lived during the preceding 25 years.

In the 1980s, nonfinancial firms, such as AT&T, General Electric, and General Motors, entered the credit card business. By 1985, many of these companies were issuing MasterCard cards.

A relative latecomer, Discover didn't get its start until 1985, through an agreement with Sears, Roebuck & Co.[14] The Discover card was launched nationwide a year later through Dean Witter Financial Services Group. By 1987, more than 700,000 merchants accepted the Discover Card. Within just a short time, the company's cardholder base had also grown considerably. Discover's one-millionth customer, a restaurant in Delaware, signed on in 1989.

---

13   Ben Woolsey and Matt Schulz, "Credit Card Statistics, Industry Facts, Debt Statistics, 2006–2007," www.creditcards.com.
14   www.discoverfinancial.com.

Discover launched its website, DiscoverCard.com, in 1995. The use of web technology, which was beginning to become important to business in the mid-'90s, proved to be a valuable step in the company's profits and overall growth potential. By 2007, some 12 million card members had registered online through the Discover website.

Visa's global sales volume had reached $2 trillion by 2001. By the end of 2007, that figure was a whopping $3.8 trillion, and reached $4.3 trillion in 2008.[15] Visa now has offices in more than 150 countries.

MasterCard went public in 2006 with the largest initial public offering (IPO) that year. At that time, its shares sold for about $39 each. Over the course of the next two years, MasterCard's stock price peaked at $227 a share.[16] By 2007, MasterCard was already bringing in a gross purchase volume of $2.3 trillion. That same year, it processed 18.7 billion transactions.

Visa went public in 2008, initially selling 446 million shares.[17] The starting price was between $37 and $42 a share, which raised about $18.8 million for the company. This was the second-largest IPO worldwide and the most for any U.S. business.

As credit cards became more popular over the years, consumers increased their spending. In fact, consumers spend more when they use a credit card for purchases. It's convenient, there's no need to carry cash, and people become more liberal with their spending. All of this is good for business, although this ease of spending has been criticized for encouraging debt. In August 2008, *U.S. News and World Report* reported, "The average American with a credit file is responsible for $16,635 in debt, excluding mortgages, according to Experian."[18]

15   Visa, Inc., Corporate Overview, www.visa.com.
16   Roddy Boyd, "Visa Can Credit Rivals for Rich IPO: Visa's Blockbuster IPO Owes Thanks to the Success of MasterCard's IPO and to the Credit Debacle Plaguing Rivals Like American Express and Discover Financial Services," *Fortune Magazine*, 2008, republished at www.money.cnn.com, February 26, 2008.
17   Information on Visa's IPO was also part of Boyd's *Fortune Magazine* article from 2008.
18   Kimberly Palmer, "The End of Credit Card Consumerism," *U.S. News and World Report,* August 2008.

To help encourage the use of credit, the card industry encouraged spending by rewarding customers with free hotel stays and airline tickets, as well as other perks that came along with increased reliance upon the cards. To put this growth in perspective, there were about 159 million credit card-holders in 2000; 173 million in 2006; and the number is projected to reach 181 million in 2010, according to the United States Census Bureau. In 2009, there are 700 million credit cards in this country. That's more than two cards for every man, woman, and child in America.

In 2009, Visa represented about 46% of the U.S. credit card market, followed by MasterCard at 36%, American Express at 12%, and Discover at 6% (see Figure 2). Globally, Visa represents about 60% of the industry's business.[19]

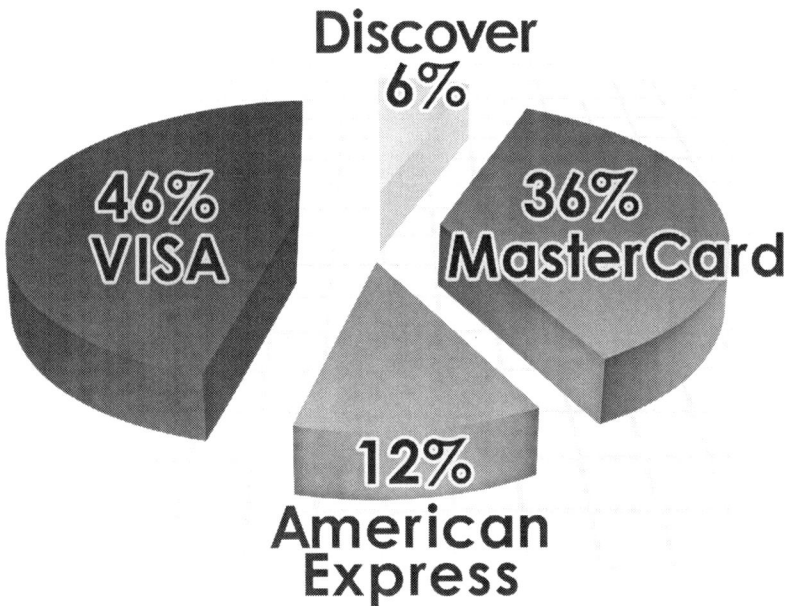

Figure 2: Key Credit Companies' Market Share

---

19   Ben Woolsey and Matt Schulz, "Credit Card Statistics, Industry Facts, Debt Statistics 2006 –2009," www.creditcards.com, November 13, 2009.

How the Industry Works

The credit industry has a number of players with basic roles in the payment process (see Figure 3). Generally, this is how the process works (depending upon the merchant processor's network system, and so on):

1. Cardholder uses a card to pay for services (in this case, it's Visa)
2. Merchant swipes the card, enters the amount, and sends an authorization request to the merchant processor
3. The merchant processor sends that request to VisaNet
4. VisaNet sends the request to the Issuer
5. Issuer approves or declines
6. VisaNet sends the issuer's response to the merchant processor
7 Merchant processor sends a response to the merchant
8. After receiving the authorization response, the merchant completes the transaction

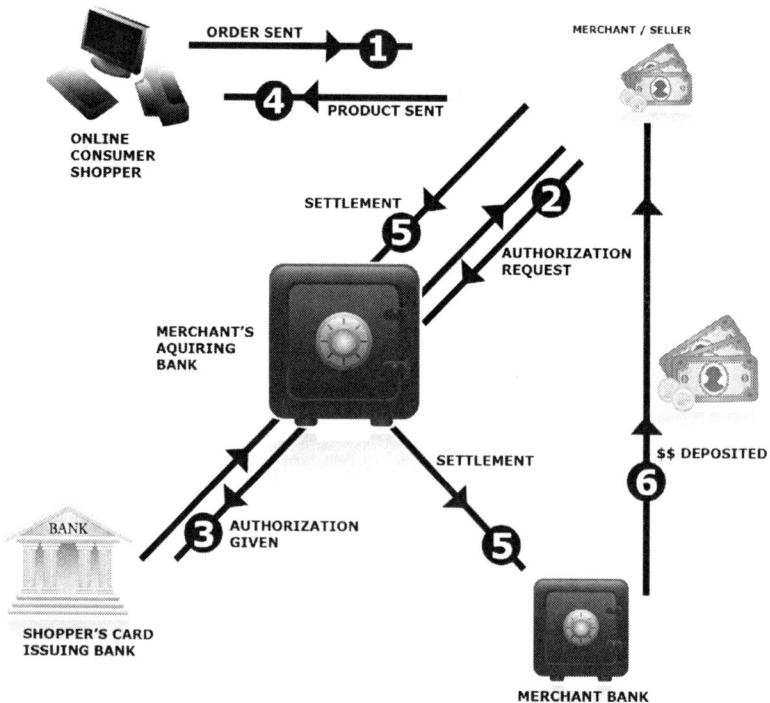

Steps in a Credit Card Transaction

A number of fees are collected as part of this process, which is one of the reasons transaction-based businesses are so profitable. When a merchant accepts a credit card for a purchase, the merchant processor must pay the cardholder's bank an "interchange" fee for every transaction that goes through the card networks (such as Visa or MasterCard). If the interchange fee is too low, then banks have a reduced profit. If it is too high, then merchants are dissatisfied.

According to an article in *Forbes Magazine*, "For each $100 a consumer spends with a Visa card, merchants cough up $2.10 in fees."[20] The rates vary based on the type of card requirements of the credit card provider. There are over 200 different card types for Visa and MasterCard alone. Business cards, for example, carry higher interchange fees to merchants than consumer cards. Other fees collected can include finance charges to cardholders for late payments, which are a highly significant source of revenue for the issuing banks.

But why would a merchant be willing to pay fees to credit card companies? Why not just require the customer to pay by cash or check?

By allowing consumers to use a Visa credit card, for example, the merchant gains an opportunity to collect money faster when consumers use their card to make purchases. Visa processes the transaction for the merchant, sparing the time-consuming and sometimes costly process of making change for cash and authorizing checks. Approved card transactions are a guarantee of payment to the merchant. By comparison, checks require a clearing process which takes several days and may not be guaranteed. Exceptions include check verification and guarantee programs such as Cross Check, where the merchant is guaranteed payment, for a fee, to the check verification/guarantee vendor. Visa's global marketing also benefits the merchant through associated name recognition and the promise to the consumer of the card's acceptance by the merchant. MasterCard does essentially the same thing.

---

20   Stephanie Fitch, "Visa, the Future of Money," *Forbes Magazine*, October 27, 2008.

Merchant fees, along with cardholder finance charges, are key sources of revenue for credit card companies like Visa and card issuers. The *Credit Card Holders' Bill of Rights*, which will go into effect in July 2010, provides a variety of reforms, such as preventing credit companies from charging unusually excessive finance charges to cardholders, making unrealistic loans to young people (including vulnerable college students, who are often away from home without the benefit of a parent's financial guidance for the first time), and closing accounts or increasing rates without sufficient notice. This legislation was developed to help consumers control their debt, especially people who face high interest rates and significant late-payment charges. Many consumers are struggling to keep up with escalating interest rate payments that occur when they don't pay their bills in full each month. It's a very real problem for many consumers as, according to a TransUnion survey in December 2008, the average total bankcard debt per borrower in the United States was $5,710.[21]

This credit card legislation also addresses other key issues, such as giving consumers more time to pay credit card bills before they become delinquent. This is important, because the average fee for late payments can be almost $26.[22] The impact of this legislation will be discussed in detail in Chapter 3.

The Infrastructure

Electronic Funds transfer (EFT) networks are the backbone of the credit card system. These networks began when automated teller machines (ATMs) were introduced in the mid-1960s.[23] While it seems almost unnatural now to wait in line to conduct transactions, especially when you can go to an ATM to get cash, or conduct transaction-based business

21   Ben Woolsey and Matt Schultz, "Credit Card Statistics, Industry Facts, Debt Statistics 2006-2009," www.creditcard.com, September 10, 2009.
22   Ibid.
23   Stan Sienkiewicz, The Evolution of EFT Networks from ATMs to New On-Line Debit Payment Products, papers.ssrn.com/sol3/papers.cfm?abstract_id=927473, April 2002.

online, there was a time when people were initially reluctant to use ATMs. Bank customers were used to dealing with people, not interacting with machines. Consumers had to be educated on the value of using debit cards to conduct transactions. They adapted to the machines rather quickly; just as many people are now paying their bills online instead of mailing out checks. Popular EFT networks include STAR, Mac, Cirrus, NYCE, and others.

EFT Definition

Electronic funds transfer (EFT) refers to the computer-based systems used to perform financial transactions electronically.

The term is used for a number of different concepts:

- **Cardholder-initiated transactions,** where a cardholder makes use of a payment card

- **Direct deposit payroll payments** for a business to its employees, possibly via a payroll services company

- **Direct debit payments** from a customer to a business, where the transaction is initiated by the business with customer permission

- **Electronic bill payment** in online banking, which may be delivered by EFT or paper check

- **Transactions involving stored value of electronic money,** possibly in a private currency

- **Wire transfer** via an international banking network (generally carries a higher fee)

- Electronic Benefit Transfer [24]

Risk-Free Income for Credit Card Companies

Visa and the other card associations, such as MasterCard, have been using the United States as a testing ground for their lucrative payment network. Although credit cards are used internationally, about 60% of Visa's business volume is in the United States.

For example, Visa doesn't have to take risks with the people who use cards with their logo — the issuing banks do. When Visa makes an agreement with a bank, it authorizes the bank to market and issue debit, credit and prepaid cards with the Visa logo. Visa then earns money on transactions made with every Visa card. When a merchant signs an agreement with a payment processor to accept Visa, part of the fees it pays would go to Visa. The same process model also applies to MasterCard and other card companies. The acquiring bank (sometimes also the payment processor) enables a merchant to accept card payments, so it takes the risk of merchant fraud. The issuing bank carries the risk of cardholder fraud or credit card payment default. Debit cardholders carry the risk because their card is tied to their bank account. (Prepaid cards would not be affected in this case for obvious reasons.) Banks earn revenue from both merchants and customers. Issuing banks earn finance charges from cardholders and banks acting as payment processors earn fees from merchants

Here's how the process works: Let's say Visa makes a deal with Bank of America for $100 credit. Bank of America issues a Visa credit card with a $100 credit line to its customer, Mr. Jensen. Mr. Jensen uses his Visa card

---

24   "Electronic Funds Transfer," en.wikipedia.org, 2009.

and spends $100 for an online purchase. Visa earns a small percentage of that sale in interchange fees.

Visa is a silent player, performing transactions in the background. Using the example above, Bank of America would pay the merchant the total of Mr. Jensen's purchase. The merchant is then charged fees by its payment processor. If Mr. Jensen doesn't repay Bank of America, Bank of America must try to recover its costs from him. Visa loses nothing on the transaction, even if Mr. Jensen defaults on his debt.

Visa and MasterCard have a great opportunity to make money. They get paid by the issuing bank for providing the platform, and they receive a transaction fee from the merchant. Even when cardholders use a Visa or MasterCard debit card, they still get a piece of the action. This is significant because debit cards will continue to become more popular with consumers at the expense of credit cards.

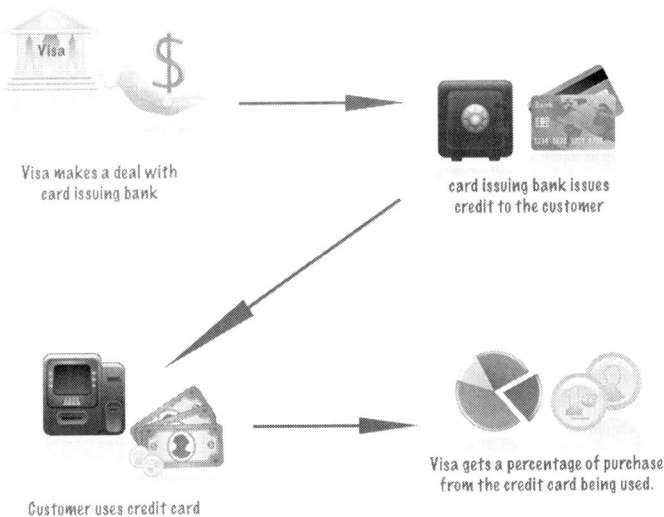

Visa makes a deal with card issuing bank

card issuing bank issues credit to the customer

Customer uses credit card

Visa gets a percentage of purchase from the credit card being used.

Figure 4: How Visa Makes Money

Visa, MasterCard and other card companies are contributing to the economy by bringing in transaction-based revenue. This revenue will continue to be replenished, because it's not based on natural resources, like the big revenue generators of some other countries. For Saudi Arabia, that revenue generator is oil. What will happen when Saudi Arabia runs out of oil in the future? The Saudis will need to look at other business opportunities. Visa and MasterCard, on the other hand, will not run out of resources.

When customers use Visa or MasterCard to make a transaction, the companies make money, and their systems grow stronger. We have to embrace this evolution of money, from paying with cash, to credit and debit purchases, to other forms of electronic transactions. The way we spend money is changing and shaping the economy — one transaction at a time. This will continue to happen, not just by swiping a credit card or paying online, but also through innovative technology for capturing payments, such as mobile and contactless devices, and disc-shaped devices used to scan payments. These devices will be reviewed in more detail in Chapter 5.

Card-Issuing Banks

The card-issuing banks are the institutions that actually give credit cards to consumers and enter into contracts with them. There are two types of issuers. Some companies, such as American Express, issue their own cards. Others are banks that contract with a credit card company, such as Visa. For example, Visa has a contractual relationship with Citibank, which allows Citibank to issue Visa credit cards. Relationships like these are very common throughout the industry. The issuers (banks issuing cards to consumers) pay a fee for the right to issue credit cards bearing Visa's (or MasterCard's, etc.) name. This fee amounts to between 1.5 to 2.5% of the bank's card association revenues, depending on its size.

Issuing banks are unique. Each may have products with different fees, incentives, charges, and credit lines. By offering so many different types of cards and related payment devices, the issuers are able to reach a broad spectrum of consumers with their products. When a payment is being processed, the credit card company sends an authorization to the issuing bank, which is identified by the credit card number. The issuing bank then transfers the money on behalf of the cardholder to the acquiring bank (the bank handling transactions on behalf of the merchant).

The interchange fee is deducted from the total amount before it is transferred from the issuing bank to the acquiring bank. The interchange fee is the portion of money that the acquiring banks have to pay to the issuing banks (most of an issuer's revenues come from finance charges imposed on cardholders, not these fees. Other sources of income include fees related to penalties, cash advances, and annual fees.)

The top five issuing banks of general purpose credit cards based on outstanding cardholder debt are:

1.   Chase

2.   Bank of America

3.   Citi

4.   American Express

5.   Capital One[25]

Card issuers may handle more than one card. Citigroup, for example, issues Visa and MasterCard, and also owns Diners Club and Carte Blanche. Citigroup was considered an innovator in the 1980s, when it introduced a new concept for tying credits to airline miles with the American AAdvantage card.

---

25   Ben Woolsey and Matt Schulz, "Credit Card Statistics, Industry Facts, Debt Statistics," www.creditcards.com, August 2, 2009.

Acquirers

An acquirer is a Visa/MasterCard affiliated bank or bank/processor alliance and is focused on the business of processing credit card transactions. The acquirers look for merchants who will accept their credit cards. The merchants may be signed up directly by the acquiring bank or through an agent or third party. "Some acquirers are banks; some are not. Some, but not all are also processors; all processors, however, are not acquirers,"[26] according to *Bankcard Today.*

Credit card companies that issue their own cards maintain control over the actual acquiring process. Even so, they sometimes use third parties to handle the work. A merchant account is the contract that is formed between the acquiring bank, sometimes through a third-party payment processor (see the section that follows on Independent Sales Organizations), and the merchant. Acquirers also assist merchants with processing services and providing support services. The acquiring banks are also forced to carry the risks of a potential transaction reversal, known as a chargeback. (See Chapter 4)

Top Seven Bankcard Acquirers – 2008 Estimates[27]

1.   First Data and Bank Alliance Partners: $1.4 trillion

2.   Bank of America Merchant Services: $295 billion

3.   Elavon: $240 billion

4.   Fifth Third Processing Solutions: $192 billion

5.   Global Payments: $119 billion

6.   Heartland: $77.2 billion

7.   First National Merchant Solutions: $48.2 billion

---

26   "An Overview," *Bankcard Today,* October 28, 2008, page 7.
27   "Sizing the Market," *Bankcard Today,* October 28, 2008, page 11.

Independent Sales Organizations

Independent sales organizations (ISOs), also called payment processors, and sales agents perform work similar to that of the acquiring banks. Their main role is to acquire new merchant accounts.

Some of the responsibilities of an ISO include setting up new accounts for those merchants to be able to process card payments, explaining to the merchant the terms and rates, and providing customer service once the merchant account has been opened. ISOs either receive fees or a certain percentage of the amount of sales in exchange for their work.

Merchants

Merchants and their customers are the foundation of the credit card industry. The merchant pays a fee, also known as the merchant discount, to the acquiring bank for accepting credit cards. The merchant discount covers the interchange fee and other costs that the acquirer or processor demands. Merchants can offer customers shopping discounts for using cash or checks, but aside from certain exceptions, they cannot charge cardholders extra (sometimes known as a surcharge) for using credit cards.

The exceptions include *convenience fees,* which some of the card companies have allowed for merchants in certain categories (i.e. government). There are other exceptions within that category as well, and each card company has its own set of guidelines. Generally, convenience fees cannot be assessed in a face-to-face environment, but if they are, there must be a separate line for those transactions; if a merchant accepts more than one payment including Visa, the merchant can only charge a flat fee, while MasterCard allows a percentage or a flat fee. Merchants accepting more than one card type must adhere to the strictest guidelines. Other exceptions include third-party collectors (i.e., Official Payments charges a fee for

IRS tax payments using a credit card) and online merchants who charge *processing fees* (i.e., Ticketmaster). Merchants with questions in this category should contact their processor.

Another exception relates to debit cards. Merchants can charge a fee to debit cardholders for PIN-based transactions, since the debit network is separate and not under the operating guidelines of the card companies.

Merchants cannot, however, in any case, impose a minimum or maximum purchase amount when customers are paying by card. You may have seen at least one of these signs, such as a sign that read "$10.00 minimum purchase on credit cards". If a merchant does impose a minimum or maximum limit, the processor, bank, or card association could terminate the merchant agreement.

As you can see, it gets pretty complicated if merchants want to surcharge customers for using credit cards. Merchants need to accept that these fees are a cost of doing business, similar to the banking fees that were imposed in the past. The last thing they should do is deter customers from using cards. Instead, they should offer discounts for using cash, which is acceptable under card operating guidelines.

Merchants also need to be aware of — and be sure they are abiding by — the operating guidelines for each card brand they accept. A merchant can be fined or have their contract terminated for violating card company guidelines.

Internet Players

Internet sales would be abysmal if it weren't for payment cards. In 2009, about 80% of online retail purchases are expected to be made with credit cards or debit cards. Email payment industry leaders, such as PayPal, represent about 9% of the market. Prepaid cards and private label cards make up about 10%.[28]

---

28   "How Do Online Buyers Pay?" www.emarketer.com, January 14, 2009.

Internet businesses are replacing many of the traditional brick-and-mortar companies. While there will always be a need for retail businesses (people want to go to restaurants or may enjoy the experience of shopping at a mall), the convenience of the Internet, combined with the ease of conducting business online, will continue to expand and take away market share from traditional brick and mortar retail businesses. Last year, industry analyst Ed Kim wrote, "*Ecommerce Journal* estimated that in 2007 about 13% of total US retail sales, excluding cards and groceries, are processed online."[29] This may not seem like a lot, but it will continue to grow. To put this claim in perspective, consider that just 10 years ago online sales did not even exist.

Think about the implications of e-commerce and how this has changed the way people do business. People used to go to their local stores to order flowers; now they are more likely to order flowers online. They used to use a travel agent to buy airline tickets; now they are probably going to book the reservations themselves online. Even holiday shopping is made so much easier with the click of a mouse. As more transactions move online, the role of the Internet players, as well as companies involved in mobile commerce (see Chapter 5), will continue to expand.

Telecommunications Companies

Phone companies are looking at ways to benefit financially from the credit card transactions that take place over their networks. AT&T, Sprint, Verizon, and others have moved into this space and are claiming a small part of the credit card companies' revenue. The phone companies "own the highway," and yet the banks and merchant service providers are driving on their road and making the most money without paying a toll.[30]

It will be interesting to see how the phone companies pursue a greater portion of this lucrative business, particularly as more transactions move to

29   Ed Kim, "Visa, MasterCard Risk Ramped up Competition," www.seekingalpha.com, March 27, 2008.
30   Jay MacDonald, "What Will Credit Cards Look Like in 25, 50, and 100 Years?" www.creditcards.com, February 17, 2009.

mobile devices. Smartphones, such as the iPhone, Palm Treo, BlackBerry, and Sprint Moto, are convenient, fast, and secure, and can ultimately eliminate the need for people to carry credit cards. The businesses that stand to make significant profits from mobile commerce are most likely the payment processors. In some cases, the processors have teamed with telecommunications providers.

Sprint rolled out a mobile payment service for small business that allows mobile phones to accept credit card payments. Sprint charges monthly fees and a set-up fee for this service, along with competitive transaction pricing.[31] Verizon and Chase Paymentech are working together to provide merchant processing with Verizon Merchant Services. They offer discounted processing fees, wireless terminal options, rapid transaction processing, and other services.[32]

Smartphones allow merchants to use mobile payment processing software applications to take payments and transmit them without having to use magnetic strip readers or other technologies. This helps push business-related functions to smartphones, and makes them a convenient way to accept card payments. At some point, smartphones may evolve to the point where they will replace laptops and credit card terminals for these purposes.[33]

Customers

Where would the credit business be without its customers? They will shape the industry with their desire to purchase items when they want to, how they want to, and where they want to. Consumers will increasingly move transactions online, not only via their personal computers, but also by using their cell phones as mobile payment devices. The use of mobile devices will increase as the ease of making mobile purchases improves and when consumers feel more confident that security issues have been addressed.

---

31   "Sprint Processes GoPayment for SMBS," www.vpico.com, May 21, 2009.
32   "Save Money and Speed Up Credit Card Transaction Times," www.smallbusiness.verizon.com,
33   Amy Buttell, "Merchants Eye Mobile Phones to Transact Card Payments," www.creditcards.com, March 17, 2009.

Chapter 2 - The Psychology of the Consumer – How We Got Where We Are Today

The consumer is the foundation of the payment card business, and the industry has spent vast sums of money trying to attract the right customers. The ideal customer is someone who pays his bill on time, but doesn't always pay it off at once. "In fact, 90% of all credit card users don't pay off their balance in full at least once a year and more than 45% of users revolve it every month," according to Meredith Whitney, writing in *The Wall Street Journal*.[34] Consumers spend considerably more when they use their cards; the average adult has about nine credit cards; and one in six families with credit cards pays the minimum amount each month.[35] There are more than 450 million Visa cards[36] in the United States alone, yet there were about 306 million people in this country as of January 2009, according to *U.S. News and World Report*.

Americans' revolving debt totaled $961.3 billion in December 2008. Non-revolving credit reached $1.6 trillion in January 2009.[37] About $879 billion of that total was credit card debt.

Debt will continue to grow as unemployment increases. In the opinion of credit card expert, Jeremy M. Simon, "The quicker we shed jobs, the faster we're going to see credit card balances increase."[38] Historically, credit card defaults are about 1% above the unemployment rate. So, defaults could rise to 11% in 2009 with a 10% unemployment rate.

Consumer debt has escalated dramatically in less than 10 years, with $2.7 trillion in debt in 1987 and $10.2 trillion in 2008 (see Figure 5). In the United

---

34   Meredith Whitney, "Credit Cards Are the Next Credit Crunch," www.wsj.com, March 11, 2009.
35   Jeremy Simon, "Credit Card Balances Jump as Economic Woes Deepen," www.creditcards.com, March 6, 2009.
36   Ben Woolsey and Matt Schulz, "Credit Card Statistics, Industry Facts, Debt Statistics," www.creditcards.com, August 9, 2009.
37   Jeremy Simon, "Consumer Credit Card Balances Jump in January," Fed Says, www.creditcards.com, March 6, 2009.
38   Jeremy Simon, "Credit Card Balances Jump as Economic Woes Deepen," www.creditcard.com, March 6, 2009.

States, in January of 2009, the average credit card balance available for consumers to borrow against was $1,673, according to CreditCards.com.[39] On top of that, the average U.S. household had approximately $8,700 credit card debt as of 2008.[40] According to Money-Zine.com, each year, between two and two and a half million people living in the United States use the services of a credit counselor. Of all of those, 75% have credit card debt.[41]

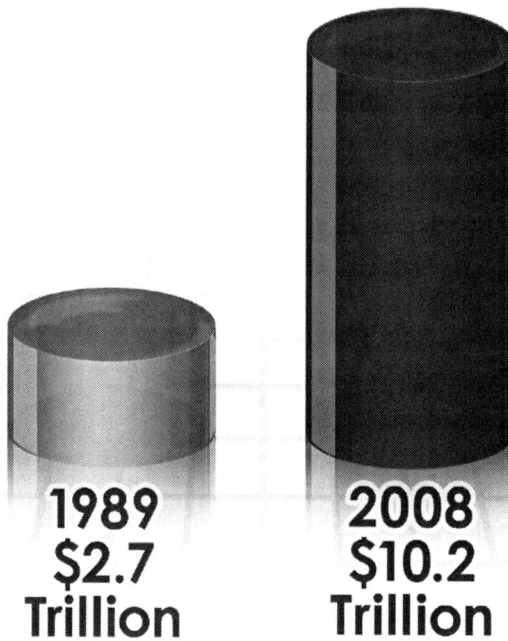

1989
$2.7
Trillion

2008
$10.2
Trillion

Figure 5: *U.S. Consumer Debt*

Generational Shifts in Spending

How did this emphasis on consumption become so important in America? One way to analyze this phenomenon is to examine the spending habits

---

39   Ben Woolsey and Matt Schulz, "Credit Card Statistics, Industry Facts, Debt Statistics 2006 —
2009," www.creditcards.com, January 9, 2009.
40   "Credit Card Debt Statistics," www.money-zine.com.
41   Ibid.

of various generations and the impact on their offspring, combined with the ever-expanding influence of the media.

Baby Boomers Enter the Scene

The Baby Boomers, who were born between 1945 and 1964, were raised by parents who may have remembered the tough times of World War II, or even as far back as the Great Depression of the 1930s.

While the early Boomers were often the first generation in their family to graduate from college, they were likely to have been influenced by the hardships of their parents. The oldest of the Boomers grew up in a more restrictive, inflexible era — a time when parents waited until they had cash before making any purchases other than their homes. Women stayed at home, and families shared just one car. The Boomers' parents often remained with the same employer throughout their entire careers and retired with pensions. Their financial requirements were predictable.

As Boomers became adults, they entered an age that allowed them more personal freedoms, and they often challenged the values of their parents. There were more opportunities for women to go to college and choose rewarding careers. Two-income families became the norm. One television was no longer enough; neither was one car per family. Designer purses and shoes became new status symbols, along with pricey cars that conveyed the appearance of success. What started as three television networks with advertising expanded to many stations, all promoting the concept of spend, spend, spend. People were inundated with messaging that encouraged consumption, and they embraced it.

Instant Gratification

With both parents working and less time available to spend with their kids, Boomers learned that it was easy to make a child happy with a new toy or treat. Busy parents also helped fuel the fast-food industry. After all, who has time to cook? Vacations became shorter, families spent more time in front of the television set, and buying more material goods just seemed like the right thing to do. Video games and television shows became the new babysitters, so the Boomers' children, Generations X and Y, watched endless hours of commercials extolling the virtues of the latest toys and products — all helping to extend the insatiable appetite of consumption. Boomers, who rebelled against the generally frugal culture of their parents, wanted to give their children the best life that money could buy. The result? The '80s became known as the "Me Generation" years.

Credit cards made these dreams a reality and provided the instant gratification that Boomers desired. Instead of waiting until they could pay for purchases, all they had to do was take out a card and enjoy the benefits of their expenditures now. Why wait? Offers for new credit cards started showing up in mailboxes with increasing frequency, often on a weekly basis. After the technology implosion of 2001, credit card issuers went after indebted borrowers by enticing them with zero interest teaser rates. The teaser rates seemed too good to be true, and they were.

As housing values rose, many people responded to the loss of jobs on the pressures of consumerism by taking money from their home equity. This just increased indebtedness, as more people used their homes as personal piggy banks. That scenario came to a dramatic close at the end of 2008: Having leveraged the inflated values of their homes to get money to spend, many people were left with upside-down mortgages when housing values crashed.

By the time Generation X'ers graduated from college (sometime between 1983 and 2006), they were often fully accepting of a culture based on consumerism. Today, 41% of U.S. college students have credit cards, and by the time they graduate college, their average student debt will be about $20,000.[42] Why save, when a card will make it possible to travel, buy the

---

[42]   Ben Woolsey and Matt Schulz, "Credit Card Statistics, Industry Facts, Debt Statistics, 2006–2009," www.creditcards.com, September 10, 2009.

latest audio equipment, get tickets to a rock concert, and drink a latte at Starbucks every day?

When choosing a career, Gen Xers were more likely than their parents to study areas that would help them get into the most lucrative businesses. They had grown up to become super consumers — watching the stock market soar to record heights in 2007, seeing home values continue to escalate for a period, and experiencing the rapid growth of the Internet and Web 2.0. They were the ideal consumers, products of their parents' own consumerism.

Rethinking Consumption

This party didn't last forever. The economic turbulence in 2008 and the slowdown in 2009 may force Gen X — and their younger cousins from Gen Y — to dramatically rethink their approach to consumption, borrowing, and credit. If they have been relying on their parents as their personal ATMs, they may be shocked to discover that Mom and Dad are no longer in a position to help. Their parents now must deal with disappearing 401(k)'s, drastically diminished home values, and more limited employment prospects as they grow older. Gen Y will see the tightening of credit and will likely be offered much less credit than Gen X. Plus, if they don't pay their bills on time each month, they will still face high interest rates and fees, even though legislation is in place to help consumers control these costs.

Ultimately, Gen Y, Gen X, and their parents will need to take a different approach to debt and credit — one that is more responsible than in recent years. Based on recent spending trends, this is already happening; consumers are starting to save again, to reduce their credit card spending, and they are increasingly using debit cards instead of credit cards. People are also beginning to defer purchases of big-tickets items. In fact, car sales are at their lowest levels in a number of years, even with record-breaking low interest rates.

The places people shop are changing, too, as more consumers today are shopping at discount stores. By mid-2009, many discount stores were seeing an increase in business, while Sharper Image, a high-end store, went out of business in 2008. Mirroring this trend, the movie *Confessions of a Shopaholic* addressed the end of the era of conspicuous consumption. It told the story of a woman who lived beyond her means and amassed a heavy burden of credit card debt. In this modern-day Hollywood fairytale, the credit card junkie became a financial advice columnist, paid off her credit cards, and lived happily ever after within a budget. What a concept!

So how are Americans living within their means? More people are turning to debit and check cards to help manage their expenses. There's a downside, however; if debit cards are stolen, they may end up costing cardholders more in fees (compared to credit cards) because there are higher liability-related deductibles for debit cards. With a credit card, the maximum you might pay if it's used illegally is $50. The maximum deductible for debit cards is generally higher. Still, as a means to more effectively manage cash flow, debit cards are gaining greater acceptance. Think of debit cards as using checks without the paperwork. Ultimately, the economic problems people are experiencing today should be a wake-up call to consumers and lead to more responsible spending.

Because of tightening credit, consumers are increasingly likely to shift their spending habits to focus more on what they *need* than what they merely *want*. They may *need* to buy a car for transportation to work. Perhaps they can only afford a Saturn, but they may *want* a Mercedes. If a Mercedes is way out of line with the consumer's budget, that person will begin to think more seriously about purchasing something affordable. A small segment of the population is even moving the clock back and turning to layaways and installment plans as one alternative to credit cards. For example, Elayway.com, an online service that brings the installment plan back in today's economy, is seeing a rise in business.[43]

Taking a Practical View

---

43   Eric Dash, "The Last Temptation of Plastic," www.nytimes.com, December 7, 2008.

Banks are no longer giving out loans as freely as they were in the pre-bailout environment. Credit card lenders estimate that they will have to write off $395 billion in bad loans over the next five years, compared with a total of loss of $275 billion in the last five years, according to *The Nilson Report*.[44] To help reduce losses, the card issuers are now more likely to work out payment plans with consumers than they had been in the past.

Overall, however, consumers won't generally have the same level of freedom and options that were once available to them with credit cards. Instead, they will be forced to take a more practical look at debt. That new perspective will ultimately help strengthen the credit card business, because it will be based on more responsible consumer spending with reduced risks. But to make this scenario a reality, credit card companies and consumer organizations must place more attention on educating people about the advantages and pitfalls of credit cards.

Schools should teach students how to use credit wisely. A formal program that makes students aware of basic credit principles related to interest, fees, credit scores, debt, and so on, would go a long way toward creating a generation that is more financially responsible.

After all, we don't let would-be drivers get behind the wheel without completing a driver's training course and passing both a written test and a driving test. So, why don't schools make understanding credit and finances a priority? Taking that step would go a long way toward helping consumers make wiser, more well-informed decisions related to credit.

The Rise of Electronic Purchasing

As young consumers move into adulthood, the more likely they are not only to have credit cards, but also to use them for a majority of purchases. For young adults, the average household spends nearly 24% of its income

---

44   Eric Dash, "Credit Card Companies Willing to Deal Over Debt," www.nytimes.com, January 3, 2009.

on debt payments.[45] By the end of 2008, the total amount of debt held by American credit card consumers was nearly $973 billion, up 1.12% from 2007, according to the April 2009 *Nilson Report*.[46] This amount includes general purpose cards and private label cards.

As consumers become more concerned about controlling debt, they will be more likely to make purchases with debit cards and prepaid cards, such as the gift cards that are readily available at supermarkets. They can put cash on prepaid gift cards and reload them when needed. Prepaid cards also open up new markets to people who don't have bank accounts, in addition to providing another way for people to make purchases without going into debt. So, a new breed of consumers may emerge in 2009 and beyond — those who are more cautious, frugal, and responsible.

While some people may think that the reduced spending by today's consumer will have a negative impact on the economy, in the longer term this will not be the case. Why? Because consumers are motivated by convenience and that will encourage spending. Cash is final, so it's more difficult to return items if you paid cash and don't have a receipt. In fact, cash is only used in about one-fifth of all transactions.

There are other disadvantages to using cash. There's no float on cash, meaning there's no lag time between making the purchase and surrendering the funds, as there is with checks and credit cards. There are no incentive rewards, such as airline miles, for using cash. Plus, a lot of people are uncomfortable carrying much cash in their wallets. Credit transactions can also be accomplished faster than cash transactions — consumers can quickly order items online instead of taking the time to go to a store and pay in person. Additionally, cash transactions can only be conducted in person.

For brick and mortar merchants, credit cards are also better than checks, because processing cards is so much more efficient. They can complete and

---

45   Ben Woolsey and Matt Schulz, "Credit Card Statistics, Industry Facts, Debt Statistics, 2006-2009," www.creditcards.com, September 10, 2009.
46   Ibid.

authorize a transaction in less time and get customers through their lines faster when customers use cards or mobile devices to pay for purchases. Of course, without credit cards, or payment methods such as PayPal, it would be more difficult for consumers to conduct online transactions.

The use of debit cards will also grow as it becomes more difficult for consumers to get credit. Reporting in the *New York Times*, Eric Dash writes, "In 2007, Americans used a debit card in roughly 21% of all transactions; credit cards accounted for about 19%, according to *The Nilson Report*, "By 2012, debit card use is expected to rise to about 29% of all transactions, while credit card use would stay about the same."[47] Internet-based transactions will also rise, with more consumers becoming comfortable using cards for online purchases and brick and mortar merchants either going out of business or moving their businesses online.

Obviously, electronic payments will continue to grow, while payments with checks and cash will slow down. When was the last time you purchased airline tickets, or hotel reservations through a travel agent? When was the last time you went directly to a box office, instead of online, to purchase tickets to a play or concert? If you have ever sent gifts to someone out of state during the holidays, didn't you notice that purchasing online was so convenient? By purchasing online, you can buy a gift without having to visit a store. You can also avoid the long lines at the post office by having the gifts shipped directly to the recipient.

Since the payment card industry earns revenue not only on the amount of money you spend, but also per transaction, growth in the volume of Internet-based transactions will help offset reductions in spending because each transaction generates revenue for the industry. Transactions made with payment cards have an additional advantage — to both merchants and consumers — of faster processing than cash and checks.

As stricter security measures are put in place to decrease the chances of credit card fraud, these transactions will be even more profitable to

---

47   Eric Dash, "The Last Temptation of Plastic," www.nytimes.com, December 7, 2008.

ıts and the payment industry. For example, during the first quar-
.009, when so many companies in the United States were losing
mu... ɟ, Visa revenue continued to grow and the company attributed much
of its profit to savings that were based on electronic processing, which
leveraged online resources instead of paper-based records.

Advantages of the Payment Card Industry

Electronic processing is also a more environmentally responsible way of
doing business. At a time when companies and individuals are looking
for ways to reduce the amount of paper consumed, electronic payments
eliminate not only the paper itself, but also the mailing costs of sending
statements to customers. They eliminate the need for fuel used by — and
the CO_2 emissions from — the planes, trucks, and trains that transport
these tons of invoices and payments each year. Quite simply, a very positive side effect of electronic processing is that it is "green." By being green,
you also don't need huge file cabinets to store paper documentation. A
simple file on a computer will do.

A side benefit from electronic processing is that you don't need to have
sensitive financial information, such as your statements, delivered to your
mailbox. You don't need to worry that these statements will get lost in
the mail or arrive too late to be paid on time. You also don't need to be
concerned about the dangers of identity theft from someone stealing your
paper statements.

The payment card industry will also profit from mobile transactions and
smart cards. It will eventually become more popular in the United States
to charge items by simply waving a cell phone, device, or smart card over
a contactless device reader than to use a credit card terminal. These
technologies (discussed in more detail in Chapter 5) will offer more convenience for customers and help expedite the payment process. Using
mobile technologies to make payments is very popular outside the United
States, particularly in Japan. It works well in countries that have not already
invested heavily in the infrastructure to support credit card payments.

Chapter 3 - Legal Issues that Will Change the Credit Landscape

While this chapter focuses on the most recent legislation — the *Credit Cardholders' Bill of Rights* — there have been numerous other legal battles, such as antitrust suits against Visa and MasterCard.

Looking back, in 1979 National Bancard Corporation brought a suit against Visa.[48] National Bancard Corporation claimed that the interchange fees Visa imposed went against the *Sherman Antitrust Act* because they constituted price fixing. The courts sided with Visa, in the end, and the practice continued.

Different case results have ended up helping different parties. According to David S. Evans and Richard Schmalensee in *Paying With Plastic: The Digital Revolution in Buying and Borrowing,* in the 1970s, major credit card companies came under scrutiny for prohibiting merchants from accepting more than one type of card, (i.e. Visa).

An anti-trust case filed in 2004 by American Express against Visa and MasterCard finally came to a conclusion in 2008. American Express contended that the other two companies stopped it "from doing bank-issued card business" in the United States. American Express reached settlements with Visa and MasterCard. First, Visa agreed to pay American Express $2.25 billion. Next, MasterCard agreed to pay American Express $1.8 billion. The combined amounts from Visa and MasterCard represented the greatest anti-trust settlement in United States history.

In 2004, Discover also filed an anti-trust lawsuit in the United States Supreme Court against Visa and MasterCard for "blocking access to the bank-issued card market in the US." The courts agreed with Discover and said that Visa and MasterCard violated the law by not letting their member banks issue cards for other companies. Discover wanted to be paid $6 billion, but settled the case in 2008 with MasterCard taking a net after-

---

48   David Evans and Richard Schmaensee, "Paying with Plastic: The Digital Revolution in Buying and Borrowing," MIT Press, 2005.

tax charge of $515 million. Visa also provided a settlement to end the lawsuit.

Changes in the bankruptcy laws impacting credit card companies in 2007 also helped to shape the legal climate for the credit card industry. The changes gave consumers incentives to manage their debt and control spending. Most households filing for bankruptcy have to file under Chapter 13, which requires that they have a plan to pay off at least some of the debt. In the past, when consumers filed for bankruptcy under Chapter 7, they could cancel outstanding unsecured debt, such as credit cards. This new regulation helped to prevent them from walking away from their bills without having to face the consequences of their actions.[49]

With consumers struggling, banks failing, and interest rates on home loans under 5%, the lowest rate in years, it is no wonder that the credit card companies have come under intense scrutiny for some of the high interest rates on revolving debt and short-lived teaser rates. These kinds of charges helped pave the way for the most striking credit reform policies in decades — the *Credit Cardholders' Bill of Rights Act,* which was passed in May, 2009 and is scheduled to go into effect in July, 2010.

The legislation gained significant bipartisan support, with landslide votes of 361–64 in the house and 90–5 in the Senate. Legislators in favor of the bill say the legislation "is necessary to stop a vicious cycle: A cardholder falls behind on one bill and watches helplessly as the rate spikes on their existing balance."[50]

It's no surprise that this regulation passed. It gained support by Congress and the public. In fact, a recent poll by CreditCards.com showed that most Americans want the government to start regulating the credit card indus-try.[51] The poll was conducted in June 2008 based on about 1,000 random

---

49   "New Bankruptcy Laws," www.lendingtree.com, August 6, 2007.
50   Anne Flaherty, "Big Changes in Store for US Credit Cardholders," www.news.yahoo.com, May 20, 2009.
51   Connie Prater, "Poll: Nearly 3 in 4 Feel Need More Credit Card Regulation; Annual 'Taking Charge' Survey Says Americans Mistrust but Need Credit Cards," www.creditcards.com, August 13, 2008.

calls to consumers. Although they wanted more credit card regulation, 82% of those people surveyed still admitted that credit cards were essential.

While many people contend that greater regulation of the industry is needed — particularly related to interest rates and fees — this begs the question, "How much regulation is enough?" One of the downsides of even the most popular regulations is that, ultimately, the government might be in a position to decide which industries would be allowed to accept credit cards and which ones would not. From there, some types of businesses could be phased out of the credit business entirely. Such control threatens many principles that helped provide Americans with business opportunities.

Consumers' Rights and Responsibilities

The *Credit Cardholders' Bill of Rights* provides protection to consumers against fee and rate increases. According to Alison Vekshin, writing on Bloomberg.com, "The legislation would ban retroactive interest rate increases on existing card balances unless a consumer is more than 30 days late, requiring companies to give 45 days' notice of all rate increases, and eliminates so-called double-cycle billing."[52]

The new bill prohibits universal default closure, which happens on one card when someone is late on another bill. The legislation also allows cardholders to set their own limits on how much they want to charge on their cards, which helps them avoid overdraft fees. It will protect consumers from being surprised with common charges, such as over-the-limit fees and charges for paying by phone.[53]

Overall, the *Credit Cardholders' Bill of Rights* protects consumers from excessive interest rate hikes and finance charges, and gives them more concise information related to details about the cost of credit. After all, why

---

52   Alison Vekshin, "Democrats Propose Measure to Limit Credit Card Abuses (Update 1)," www.bloomberg.com, January 15, 2009.

53   Anne Flaherty, "Big Changes in Store for US Credit Cardholders," www.news.yahoo.com, May 20, 2009.

should you have to go through 30 pages of fine print just to figure out the terms of your agreement with the credit card company? Who reads those pages?

To cover losses associated with reduced interest rates, credit card companies may charge extra by increasing the annual fees for owning a card. They may also attempt to switch fixed-rate cards over to cards with variable rates. That's because, after the new legislation goes into effect, lenders will still be able to raise variable interest rates, which are tied to indexes, typically the prime rate or the LIBOR.[54]

The bill also helps prevent young people from being tempted by the allure of easy credit. People under 21 will need to prove that they can repay the money they borrow, or have a parent or guardian agree to pay off the debt if they default. Other restrictions also prohibit creditors from extending credit to consumers under 18, except under special circumstances.

Increasing industry oversight is necessary to help curb the patterns of spiraling debt, as people get deeper and deeper into a hole trying to pay off their credit cards. The average consumer can also become overwhelmed by the fine print that contains the details related to the credit card and to their privacy. The consumer's best option is to limit or eliminate the balance each month and avoid the extra fees and interest charges.

Small Business and the Bill

While consumers benefit from the *Credit Cardholders' Bill of Rights* legislation, the bill currently does not do much to help regulate credit for small businesses, which often rely on credit cards to finance their operations. In fact, less credit will be available to them, especially to start-up businesses. This is a concern because, based on a recent survey of the National Small Business Association (NSBA), about two-thirds of overall small business debt is comprised of credit card debt. Almost all of the owners surveyed

---

54   Ellen Cannon, "The Near Credit Card Future," www.bankrate.com, March 12, 2009.

paid an interest rate of 15% or higher and over half of them had a combined credit limit of $50,000 or more. Most of the people surveyed also reported that the terms of their credit agreements had worsened over the last five years.[55]

NSBA is supporting legislation to help small businesses achieve some of the new credit card benefits that will be available to consumers in 2010. This includes prohibiting universal default, double-cycle billing, retroactive interest rate hikes, interest charges on transaction fees, and other charges.

While the approach of increased oversight of the credit card industry appears to benefit consumers, particularly those who make only minimum payments each month, one unintended consequence is that about 40% of unused credit lines could be pulled.[56] "We believe the proposed restriction is unnecessarily stringent and would severely curtail the ability of creditors to react to adverse changes in a borrower's risk characteristic during the term of the account," the Office of the Comptroller of the Currency (OCC) told the Fed in public comments.[57]

What's happening now in anticipation of the bill's enactment in 2010? Lines of credit are disappearing, and more issuers are switching fixed rates to variable rates. As Bankrate.com explains the situation, "The variable rates could rise with index rate-related increases, making them a better money maker for issuers. About 66% of credit card offers are variable."[58]

Meredith Whitney, a banking analyst at Oppenheimer and Company, predicted that credit card companies would reduce credit lines by $2 trillion, or about 45% of all available credit, by mid-2010.[59] Discover, for example,

---

55   *NSBA Small Business Credit Card Survey*, National Small Business Association, www.nsba.biz/docs/09CCSurvey.pdf, 2009,

56   Mantell, Ruth, "Tougher Rules for Credit Card Issuers in Work," i2credit. com, February 1, 2009.

57   Ibid.

58   Ellen Cannon, "The Near Credit Card Future," www.bankrate.com, March 12, 2009

59   Neha Singh and Amiteshwar Singh, "Credit-Vard Industry May Cut $2 Trillion Lines: Analyst," www.reuters.com, December 1, 2008.

closed 3 million accounts by January 2009 and is expected to close 1–3 million more for certain customers.

Under the theme "no good deed goes unpunished," some other unintended negative consequences may accompany the bill. It will cause banks to be more discerning regarding to whom they want to issue credit. Some people have recently noticed that their credit lines have been cut, even when they have paid on time. According to one source, the new requirements could cost the banking industry more than $10 billion a year in interest payments.[60]

No one is feeling sorry for the credit card industry. Those companies will find a way to make up the lost interest payments, such as by adding fees to some services that are currently free of charge. They will also become more innovative and move more aggressively toward a payment system that rewards people who make online payments, eliminating the costly and time-consuming process of handling check payments. Expect, too, to see credit card companies respond by reducing the length of time available for introductory rate offers. Instead of being offered a new card at a low interest rate for 12 months, the companies might change the offer to cover just 3 or 6 months.

Highlights of the *Credit Cardholders' Bill of Rights*

- Payments must be applied to the balance with the highest interest rate or proportionally to all balances.

- Issuers must give more notice before changing rates. Consumers will have more time to pay bills.

- Consumers will know in advance when to expect a rate increase.

---

60   "U.S. Regulators Adopted Sweeping New Credit Card Rules," www. smartcardtrends.com, December 18, 2008.

- Balances won't be computed with two-cycle billing.

- Issuers will not be allowed to unfairly add security deposits or fees for issuing credit.

- Fees for subprime cards will be more closely regulated.

- Credit for college students and consumers under 18 will be more closely restricted and regulated.

- Rate-related information will be communicated more clearly.

- Creditors must send statements 21 days before the due date of the outstanding balance.

Recommendations

Now is the best time for more self-regulation in the credit card industry, even before the new rules become effective in July 2010. Credit card companies should voluntarily, concisely, and more clearly explain the terms and conditions of their cards to consumers. Rates should be closely scrutinized to continue to provide a good profit for the credit card companies but without undue fees and hikes.

Credit card companies are taking measures to provide the right amount of credit to people based on their individual risk. Surprisingly, American Express is even offering to pay its less-valued customers $300 to close their accounts. A more conservative approach to lending will ultimately facilitate more responsible borrowing and lending. The sheer volume of card handling should help to offset any losses the credit card companies might incur from reduced fees and rates.

The leaders in this industry are resilient and resourceful. They will find ways to remain profitable, even in an environment where profits are more

Michael A. Brooks

regulated. They will incentivize consumers for using more cost-effective business practices, such as agreeing to receive statements online instead of in the mail. This approach will continue to provide substantial savings on paper, printing, and mailing costs. These savings will help to offset lost revenue from rates and fees, and provide a greener, more socially responsible way of doing business.

Chapter 4 - Security and Fraud Prevention

The biggest concern many people — from merchants to banks to card-holders — have about credit cards is related to data security. Credit and debit card theft is a growing problem. Credit card companies recognized the importance of maintaining secure transactions and joined together to create the Payment Card Industry (PCI) Security Standards Council to address security problems.[61]

Launched in 2006, the PCI Security Standards Council was founded by American Express, Discover Financial Services, JCB International, MasterCard Worldwide, and Visa, Inc. Its role is to develop security standards to protect cardholder account information. The Council's work includes improving standards, dealing with data storage issues, and implementing and disseminating the standards.

Payment Card Industry (PCI) Data Security Standards at a Glance

The PCI standards, which were updated in October 2008, include these highlights:

- Payment card industry members must have secure networks and firewalls.

- Members must take action to protect information, whether it's stored or sent through the networks.

- Members must use encryption and maintain a vulnerability management program.

---

61   Information about the PCI Security Council, its formation and the standards it sets forth can be found at www.pcisecuritystandards.org.

- Anti-virus software must be updated.

- Systems and applications must be secure and controlled.

Members of the PCI Security Council continually monitor cases of account data compromise. A security breach and subsequent compromise of payment card data affects many different entities, from cardholders to business owners, which is another reason why it is so important to have the industry closely regulated.

In spite of these regulations, breaches happen, and the impact is serious. In January 2009, Heartland Payment Systems identified a security breach due to malicious software that had apparently been installed inside the company's processing system. Heartland, which processes credit cards for about 80% of the restaurant industry, was certified as PCI compliant. PCI compliance should only be considered a starting point in the fight against fraud. The number of cardholders affected by Heartland's breach was not reported but the company processes over 100 million transactions a month for over 175,000 merchants. The stolen data may have been used to create copied versions of original cards. Heartland's breach may have impacted tens of millions of cards.[62]

In fact, this breach demonstrates how Internet-based transactions can actually be less risky than retail point-of-sale (POS) transactions, even though POS transactions collect magnetic stripe information. The Heartland breach was due to a weakness in data security in which the physical cards were compromised; it was *not* a breach of security for online credit transactions. The stolen data could have been used to create copied versions of original cards.

This was one of the largest breaches in history. Since it compromised cardholder information from the cards' magnetic stripes, many banks had to send out replacement cards and alert users to watch their statements very closely — an action that had both a high financial cost and a high risk of confusion for cardholders.

---

62   John Borland, "Malware Swipes Millions of Credit Cards," www. technologyreview.com, January 22, 2009.

As of February 2009, some 600 United States institutions were impacted by the Heartland breach, and more than 45 million cases of identify theft could have resulted if proper actions had not been taken immediately. How can you easily update thousands of retail merchants when such a breach occurs? You can't. But you can control this problem much more easily online by tracking bank identification numbers (BINs). A BIN is the first six digits of a card number that identifies the card network and issuing bank.

The fact that this breach could even occur is very disconcerting. Heartland is a top 10 processor and processes payments for a majority of the restaurants in the United States. The company charges low fees, but Heartland makes money from doing a high volume of business. The company was vulnerable based on the way it stored data. In fact, this breach was so serious that in March 2009, Visa temporarily removed Heartland from its list of providers that comply with PCI standards.

Although some people still claim that Internet transactions are less secure than those with a card-present environment, this breach demonstrated how Internet transactions can actually be safer when security measures that track BINs from online purchases are used.

Think about the impact of this breach on merchants. Most of Heartland's merchants are small to medium-sized businesses and the average merchant in its network does about $350,000 in Visa and MasterCard transactions each month. When security at a payment processor like Heartland is breached, that impacts Heartland and its merchants. Heartland may get stuck with big expenses when their merchants are the victims of fraud.[63]

Updates to PCI Compliance

In October 2008, there were a number of revisions to the *PCI Data Security Standard (DSS)* 12 requirements. Two changes are worth noting. One states that if you store data offsite, the storage locations must be visited

---

63   Jordan Robertson, "Credit Card Data Breach Could Be Among Largest," *San Jose Mercury News*, January 22, 2009.

and validated as compliant. The other change says that by June 30, 2010, users of wired equivalency privacy (WEP) — a software application that protects data that travels across wireless networks — need to upgrade to a more sophisticated process known as Wi-Fi-protected access (WPA). This requirement will help increase the security of transactions.[64]

Protecting Against Fraud

Credit card fraud occurs in a variety of ways, and it costs businesses billions of dollars each year. The American Association of Retired Persons (AARP) cautions, "Businesses pass these costs on to cardholders in the form of higher prices, interest rates, and fees."[65] While cardholders aren't usually responsible for paying for fraudulent charges, it can be time consuming to try to get these charges removed. Cardholders should look closely at their credit card statements and notify their card companies about unauthorized charges. That will help to reduce their liability in the event of fraud.

Because debit cards are similar to checks, cardholders have less protection against fraud than when they use a credit card. With debit cards, money is quickly removed from their account, so there is not usually enough time to do a stop payment, which can be done with a check. To avoid any concerns about whether a merchant will deliver undamaged merchandise, cardholders should use a credit card for that purchase.

Debit card holders should also be cautious at the checkout counter, ATM, and other places where debit cards are used. Obviously, it's important to make sure others do not watch when a PIN is being entered. In addition, cardholders should frequently review their accounts and immediately notify their bank of any discrepancies.

One common fraud technique occurs when a thief "tests" an ATM card by putting on a small charge, such as $4.95, and labeling it as a processing

---

64   "Tightening Security," *Bankcard Today*, December 2008.
65   AARP Outreach & Service, "Credit Card Fraud," www.AARP.org, www.aarp.org/money/consumer/articles/fraudscreditcards.html.

fee or something else that's innocuous. If that charge goes through, the company may try another fee of about $25 a few months later. If that charge also goes through, the perpetrator may finally put through charges of hundreds of dollars or more.

Some protections are available with debit cards, but it's important to act quickly. Cardholders must notify the bank within two days after they discover the theft. The liability for unauthorized transactions is capped at $50. Failing to notify the bank in time can cost a cardholder up to $500 or more.[66] Fraud has gotten so rampant that one security strategist mentioned how a broker, who was trying to sell credit card numbers, guaranteed that he'd replace them instantly if the numbers were bad and that he would provide discounts for large orders.[67]

The Perils of Chargebacks

Merchants can also be victims of fraud. Fraud related to chargebacks can occur even when customers dispute legitimate charges on their credit cards. This can be known as "friendly fraud," such as when a consumer uses a card, and then falsely claims he or she never received the items or services.

A chargeback usually happens within 120 days of the transaction date (most banks have a time limit on initiating chargebacks). The cardholder may claim that the charge is not recognized or the merchant never delivered the merchandise.

When a chargeback occurs, the merchant's bank account (tied to the merchant account) is debited, and the merchant must provide proof that the transaction is valid and complies with Visa/MasterCard rules and regulations.

---

66   The Debit Card Debate," *FDIC Consumer News,* www.fdic.gov/consumers/consumer/news/cnspr06/debitcard.html, Spring 2006.
67   Asavin Wattanajantra, "RSA Europe: The Growth of the Underground Hacker 'Economy,'" www.itpro.co.uk, October 28, 2008.

Common Reasons for Chargebacks

Chargebacks occur for many reasons, some of which are fraud. Others are valid complaints. The following are the primary reasons for chargebacks:

- The customer disputes the quality of the merchandise.

- A refund credit is not received.

- Proper authorization is not obtained at the time of the transaction.

- Merchandise is not received by the customer.

- Charges are duplicated.

- The consumer is perpetrating a fraud.

- Fraudulent card use.

Card companies have placed guidelines on chargebacks and how they affect the merchant's account. For example, if more than 1% of the merchant's sales result in chargebacks, Visa could shut down the merchant account. Or, if a merchant receives too many chargebacks in three consecutive months, Visa and MasterCard may also close the account or put the merchant on their blacklist. That means the merchant won't be able to accept credit cards. Even if Visa and MasterCard still let the merchant accept credit cards, the merchant could face significant fines if the chargeback-to-transaction-ratio is too high.

The key is to prevent chargebacks from happening. In the cases when they do occur, it's essential to minimize their impact by effectively fighting chargebacks. Friendly fraud is more pervasive with online businesses, because it's easier to commit fraud when a credit card is not physically swiped through a card reader and a cardholder signature is not on file.

Merchants need to pay attention to *chargeback ratios*. While Visa and MasterCard don't approve merchant accounts, they do hold all the cards when it comes to setting chargeback ratios. Chargeback ratios are calculated by dividing the merchant's monthly sales by the number of chargebacks, or the total number of chargebacks registered in any given month, or both. If a merchant has multiple merchant accounts, receiving chargebacks on a merchant account that is closed, however, does not affect a merchant's chargeback ratio.

Merchants can take proactive steps toward facilitating approval of their merchant account and receive a fair chargeback ratio, which is generally about 1%. They should provide their bank or processor with as much detailed and specific information as possible, including:

- How long their company has been in business and what they sell

- If they have processed credit cards in the past

- The dollar amount they expect to process per month

If fraud detection controls aren't strong enough, a merchant could end up with too many chargebacks, which could put the merchant on the dreaded Terminated Merchant File (TMF) list. If a merchant is on the TMF list, this means that a bank has terminated a merchant account with the merchant, and this sends up a red flag to banks that the merchant is a credit risk. Getting on the TMF list is the equivalent of getting blacklisted by the credit card companies.

Merchants should concentrate on preventing chargebacks before they happen. The chargeback process takes considerable time, during which the merchant is denied the revenue of the disputed sale. Additionally, there is a considerable amount of paperwork involved.

Steps in the Chargeback Cycle

The chargeback process has a series of formal stages:

1. **Presentment**: The merchant makes the original credit card sale and charges a customer's card.

2. **Chargeback:**

    a. **First Retrieval Chargeback**: A customer disputes a charge to his or her credit card. A retrieval request is sent to the merchant to dispute the transaction. In some cases, the retrieval request is not required, such as if the reason for the dispute is fraud.

    b. **Second Presentation or Re-presentment**: This is when the merchant has an opportunity to respond to the first chargeback retrieval request.

    c. **Second Chargeback**: If the first presentment is rejected by the cardholder, the issuing bank files a second chargeback.

3. **Arbitration**: This is when Visa/MasterCard staff review the situation. The merchant could pay up to $750 in fees, even if the merchant remains in arbitration. Note: A chargeback goes on the merchant's record.

Preventing Chargebacks

The best approach to prevent chargebacks related to fraud can be to use payment processing technology with built-in fraud prevention best practices. Some of these practices include:

- Verifying the customer's address when the transaction is processed. This is critical. MasterCard or Visa will verify that the numbers in the customer's address match the billing information.

- Requiring the customer to enter the security code on the card.

The reason for this step is that someone would most likely need to have the physical card in order to complete the transaction. A copy of a credit card bill or the account number won't do.

- Using a software solution that watches for BINs from cards originating from a suspicious geographic region. Ideally, the solution should be able to block a card in that region, even after one transaction, and prevent fraudulent charges from taking place.

- Being extra vigilant with customers who have recurring payments and have disputed several transactions which have resulted in chargebacks. For example, if a customer sets up a recurring payment for vitamins, and disputes his October order, the merchant may get a chargeback. Merchants should first contact the customer to find out what happened. If no resolution is found, the merchant should turn off the customer's recurring account to prevent future shipments and additional chargebacks...

- Providing customers with a copy of sales receipts – every time. This information should be readily available. A retrieval can be a precursor to a chargeback. Requests for the retrieval come from the issuing bank. The merchant can simplify this process by leveraging technology to supply those requests and avoid tracking down extensive paperwork.

Leveraging Technology and Services

It's important for merchants to closely monitor the percentages of charge-backs and get an automated alert, which is warning to prevent a merchant from exceeding the Visa and MasterCard chargeback thresholds. If a merchant is put on warning, it may be a few months before it loses its ability to accept credit cards. This is like a slow death for retail merchants. Even in the first month, they need a project plan to identify how they will lower their chargeback ratio. If they don't do this, they are potentially open to fines and possible termination.

Fraud is a leading cause of excessive chargebacks. Chargeback preven-tion and management solutions enable merchants to focus on delivering great customer service. Criminals are using more sophisticated tactics to bypass many of the traditional transaction management and data security checkpoints, which only confirms if a credit card is legitimate and if user-reported account information matches those on record.

It's important for merchants to use a software solution that identifies whether the purchaser is the legitimate cardholder, and then to determine each customer's potential for charging back. Through sophisticated analy-sis, the solution should be able to identify traits and patterns associated with fraudulent or high-risk orders and rank each customer using a grading scale. From there, it should be able to help determine each transaction's chargeback or fraud potential and stop the transaction from occurring.

Merchants can reduce risk and increase profitability by managing transac-tions with a comprehensive checks-and-balances system that addresses all possible angles. They can take action against friendly fraud with a software solution that addresses chargeback collections. Regardless of the outcome of the chargeback process, when a consumer decides to initi-ate a chargeback from a valid transaction, the merchant is still owed the money. A professionally managed software solution increases recoveries from chargebacks through research, documentation, and fast responses. Vendors that provide this solution should also provide supporting services, where they research every single chargeback, from order process to delivery of goods, and prepare any necessary documentation.

It's important for the merchant to have access to a comprehensive chargeback database with advanced client tracking that prevents chargebacks from occurring. This service should automatically block customers with false chargebacks from initiating charges within the merchant's entire network.

Transaction management systems are not one-size-fits-all. The solution vendor should provide an audit of the merchant's chargeback history and examine specific chargeback cases to design a custom payment processing and chargeback prevention program tailored to the merchant's unique business needs. Identifying patterns and areas of weakness can protect merchants from future chargebacks or fraudulent activity.

BIN blocking can help merchants prevent chargebacks before they happen. With this approach, BINs are automatically blocked if they have a higher chargeback potential. An automated alerting system notifies the merchant when BINs hit specified levels for volume, decline ratios, and chargeback ratios.

Chapter 5 - Trends and Growth

After a significant economic downturn in the latter half of 2008, that year's holiday shopping season turned out to be the worst since at least 1970.[68] Numerous analysts have contended that the worsening economic conditions will have an adverse impact on credit card companies. Visa's debit card business helped strengthen the company, but with unemployment now reaching double-digits, we are in uncharted territory.

Many economists are concerned that battered banks will have to deal with yet another fallout of the economic crisis — credit card defaults. In an interview with CNN's David Ellis, Bank of America CEO Ken Lewis expressed "no doubts that 2009 would be an awful year for the credit card industry."[69] With U.S. consumers having nearly $1 trillion of credit and charge card debt outstanding as of October 2008, one report by Reuters predicted that losses for the credit card industry could top $70 billion in 2009.[70]

While some analysts expressed concern that 2009 revenue growth for Visa and MasterCard would decline from the previous year, they still predicted growth — just less of it. For Visa, that means an earnings-per-share estimate of 15% in FY 2009 and 21.4% in FY 2010, instead of 18.1% and 20.6% growth.[71] While unemployment certainly impacts an individual consumer's ability to spend, it's important to keep in mind that, even when unemployment is at 10%, the other 90% of the workforce are still working and spending.

Karik Mehta, FTN Midwest Securities Corp, contends that in 2009, "Banks are clearly in a difficult environment, but the payment mechanism isn't going away. The ability to use credit and debit cards isn't going away." He points out that people use credit or debit cards for purchases they would have never used them for in the past — such as parking and fast

68   Reuters, "Losses for Credit Card Companies Could Top $70 Billion," mydigitalfc.com, January 3, 2009.
69   David Ellis, "Banks' Future Woes in One Word: Plastic," www.money.cnn.com, March 10, 2009.
70   Reuters, "Losses for Credit Card Companies Could Top $70 Billion," mydigitalfc.com, January 3, 2009.
71   "KeyBanc Lowers Estimates and Price Targets on Payment Processors MasterCard (MA) and Visa (V)," www.streetinsider.com, January 15, 2009.

foods — because the cards are convenient, and because MasterCard and Visa are "doing a very good job of encouraging customers and providing opportunities for them to use credit cards."[72]

As of January 2009, outstanding credit debt in the U.S. totaled just under $1 trillion dollars. The card companies sent 27% fewer solicitations to new cardholders in 2008 than in 2007,[73] and this trend is expected to continue for the foreseeable future.

The economic wake-up calls of 2008 and 2009 have inspired many consumers to take a deep breath before even considering whether to make a credit card purchase, especially if the expense for that purchase is significant. So, while the amount of spending is slightly less per transaction, the sheer volume of transactions is still expected to grow dramatically. There will be more ways for people to use credit cards, and consumers will make more frequent, but smaller, transactions. According to a report by the U.S. Department of Commerce, "Future growth will come from the introduction of new products in developed markets and from increased penetration of existing and new products in developing markets."[74]

The Commerce report states that growth in electronic payments has surpassed the general economic growth. For example, this situation has made it possible for people who do not have bank accounts to have the benefits of banking by leveraging prepaid cards. The report also cited the role of electronic payments in the e-commerce, travel, and tourism sectors. "Increasing the existing share of electronic payments in a country by just a margin of 10% will generate an increase of .5% in consumer spending,"[75] the Commerce report said.

---

72   David Ellis, "Bank's Future Woes in One Word: Plastic," www.money.cnn. com, March 10, 2009.
73   Ron Lieber, "Credit Card Companies Go to War Against Losses," www.nyt. com, January 31, 2009.
74   Scott Schmith, "Credit Card Market: Economic Benefits and Industry Trends," U.S. Department of Commerce, International Trade Administration, March 2008.
75   Scott Schmith, "Credit Card Market: Economic Benefits and Industry Trends," U.S. Department of Commerce, International Trade Administration, March 2008.

More Security Concerns

Data security is an even greater priority today than ever before. According to a survey by Visa, the majority of merchants had not adequately protected cardholder data. In fact, one industry survey found that only 17% of 231 merchants surveyed had followed industry guidelines for security. For example, retailers and processors kept data they should not have retained. When account numbers and personal identification numbers (PINs) are kept in unencrypted databases, the data is susceptible to hacking. When hacking of unprotected data occurs, the issuing banks' processors and merchants are subject to fines. Then the banks fine the merchants. According to a leading industry analyst, these conditions will help to force compliance, hopefully reducing or preventing data security issues. Fines for data security and compliance violations are forcing merchants and banks to implement and monitor data protection controls. Prepaid cards are one answer to consumer concerns about security because they do not contain identifying personal data.

In Europe and parts of Asia, payment cards include embedded chips. These chips provide extra security, because cardholders need to enter a PIN before using the card. Until such cards become commonplace in the United States, consumers in this country will be vulnerable to data security problems.

Limiting Rate Hikes

As the July 2010 deadline of the *Credit Cardholders' Bill of Rights* nears, banks are working to enact reforms. For merchants, the new credit card regulations schedule attempts to limit interchange rate hikes. This legislation will help make consumers more aware of possible rate increases by requiring the card issuer to provide clearer details about interest rates during the application process. Banks should take advantage of this opportunity to fully disclose information to prospective customers, instead of trying to fight against the guidelines.

Rewarding Customers

Credit card rewards programs can encourage responsible behavior. The Schwab Bank Invest First Visa credit card, for example, provides debt management and savings. It also allows consumers to put "the 2% cash back they earn on purchases into a Schwab Area Brokerage Account."[76]

Another product, Discover Motiva, encourages consumers to pay the minimum balance on time and gives them cash back for doing so. While it is good for a company to reward cardholders for paying bills on time, unfortunately, Discover is encouraging them to carry over debt simply to get a reward.

New Markets in Developing Nations

About 70% of the world's population belongs to the "unbanked" segment.[77] This represents a significant new market for payment cards in Latin American countries, such as Brazil and Mexico, as well as a number of other emerging markets.

As nations become more economically advanced, their residents are more likely to increase their use of credit cards. Countries with some of the most dramatic increases in credit card usage in recent years include Hungary, the Czech Republic, and Russia (see Figure 6). Eastern Europe may surpass India, China, and South Korea with 1% growth of electronic payments.[78] As a result, the nations of Eastern Europe will need to expand their technology infrastructure to support this growth.

---

76   Holmes, Tamara, E., "How Credit Card Rewards Can Bail You Out of Debt," www.creditcards.com, April 6, 2009.
77   Ibid.
78   Scott Schmith, "Credit Card Market: Economic Benefits and Industry Trends," U.S. Department of Commerce, International Trade Administration, March 2008.

Change in Credit Card Use in Selected Countries (1998–2005)

Country	1998	2005	Change (%)
Argentina	0.24	0.35	44
Brazil	0.14	0.38	168
Chile	0.14	0.19	37
Colombia	0.04	0.07	70
Czech Republic	0.00	0.07	2,323
Hungary	0.00	0.09	3,022
India	0.00	0.02	405
Mexico	0.06	0.13	112
Russia	0.00	0.02	>9,999
United States	1.80	2.53	40
Venezuela	0.13	0.12	−5

Figure 6

Source: Economist Intelligence Unit, *European Marketing Data and Statistics 2007* (London: Euromonitor International, 2007, *International Marketing Data and Statistics 2007,* and Visa International)

India is the fastest-growing market in the Asia Pacific region, with credit card usage growing at a rate of 30–35% a year. The features of their cards are very important to the Indian consumer, and include more benefits than cards in the United States. For example, standard cards in India include discounted medical insurance and free accident insurance. Indian consumers also get statements that break down the categories of their spending to help them budget. This information is similar to detailed reporting available with online banking and some credit cards in the United States.

The movement to get people to use credit cards represents a significant cultural shift for the Indian consumer, who has traditionally paid cash for most items. One of the challenges for growth is that India lacks a strong technology infrastructure to support this new credit card demand. Yet, because the regulatory environment is more favorable to the banking industry, this environment may help spur more investment in the electronic systems needed to support credit cards.

Indian banks could give Visa a run for its money. Indian financial entities are working to establish a transaction settlement company called India Pay, which would compete with MasterCard and Visa.[79] The banks in India support this arrangement, because they would save money with India Pay by avoiding interchange fees charged by Visa and MasterCard.

Malaysia, too, offers robust growth opportunities to the credit card industry. The number of cards issued and the amount of credit card spending doubled between 2002 and 2007. Malaysian credit card rewards programs are also very aggressive.[80]

Turkey is also expected to represent a growing market, although regulations keep interest rates low. As of July 2008, credit card interest rates in Turkey were 4.39%.[81]

One of the greatest opportunities for expansion is in China. The Olympics helped to pave the way for this market because ATMs and other services were needed to support the demands of tourists. In 2003, there were only 3 million credit cards in China. By 2008, that soared to 104 million credit cards. "In the next 5 to 10 years, analysts say, China could issue 1 billion new credit cards."[82]

American Express is targeting middle-class customers in Mexico, trying to appeal to them with its Blue Card, which also reaches younger demographics. This card offers attractive rewards, which is helping to generate more business. "Mexico is one of the 7 biggest markets in the world, and

---

79   "Financial Cards in India," www.euromonitor.com, April 2008.
80   "Robust Growth in Malaysian Credit Card Market, www.btimes.com, November 5, 2008.
81   "Report: Turkish Credit Card Market Expects Strong Growth," www. atmmarketplace.com, July 22, 2008.
82   Matt Schulz, "Credit Cards Around the World, China," www.creditcards. com, October 28, 2008

the number-one Latin American Market for American Express," according to the *Commerce Report.* [83]

Figure 6 also shows the change in credit card usage from 1998 through 2005 in selected nations. During this time, significant expansion occurred in Mexico, Columbia, and other Latin America countries. The growth in Russia, Czech Republic, Hungary, and India during that period is even more significant and is a sign of increased expansion going forward. The United States led in market penetration, with 1.8 cards per person in 1998 to 2.53 cards in 2005, as referenced in the above figure.

A Cultural Shift

Significant changes in consumer culture have paved the way for more frequent use of credit, debit, and prepaid cards. For many years, merchants only accepted cash or checks. Now, in addition to online retailers, there is a significant shift in other businesses that will take credit or debit cards. In fact, some businesses will no longer accept checks or cash. For example, airlines are discovering that purchases made in flight can be done easier and faster when no cash is involved. This practice also prevents skimming. Some vending machines now accept credit or debit only. Until recent years, people did not have the option of purchasing fast food with anything but cash; but that has changed as well, with credit card swiping allowed at both the counter and the drive-thru windows.

Prepaid Cards On the Rise

While retail businesses that require card swipes are decreasing, there will be other opportunities with companies that have not traditionally used credit, debit, or prepaid cards. One of those areas of growth includes

---

83   Scott Schmith, "Credit Card Market: Economic Benefits and Industry Trends, U.S. Department of Commerce, International Trade Administration, March 2008.

online payments for professional services, such as doctors, lawyers and accountants.

Prepaid cards are becoming increasingly attractive, not only for people who don't have bank accounts, but also for those who choose to use these cards as gifts or to make payments. Prepaid cards are particularly appealing to people who could not pass a credit check or employment verification.

Prepaid cards may also be an attractive option for businesses. One company in the United Kingdom offers a prepaid card program that makes it easy for employees to submit expenses and manage their budgets. Another advantage, particularly in Europe, is having the cards available in multiple currencies. As a result, when a business traveler goes from London to France, he doesn't need to worry about exchanging pounds for Euros.

Credit Cards for Students — A Double-Edged Sword

Some people have concerns about the number of credit cards issued to young people. Students often graduate with significant credit debt that has been far too easy for them to obtain. Yet, there are many advantages to the colleges for offering cards to this new market. Colleges receive income for providing this service. Bank of America, for example, has an agreement with 700 colleges and alumni associations. One of those schools, the University of Michigan, will receive $25 million over the course of its 11-year agreement with Bank of America in exchange for student credit card enrollments. Most of the money earned this way is used to fund scholarships.[84]

Students like the freedom offered by credit cards. With education in how to use credit wisely, students can benefit greatly by being able to use credit

---

84   Johnathan D. Glater, "Colleges Profit as Banks Market Credit Cards to Students," www.nyt.com, January 1, 2009.

cards to buy books, attend school, and pay for tuition and meals. The new *Credit Cardholders' Bill of Rights* will limit this access to credit for students. So, while the bill will protect certain rights, it does have some drawbacks.

The Changing Face of Credit Cards

Credit cards are now coming in all shapes, sizes, and formats, including stickers, coins, key fobs, phones, and more. In Paris, attendees at a conference were introduced to a car key for a BMW that could be used just like a credit card. Other options include biometric cards, based on fingerprints, and smart cards that can be swiped and provide increased security.

Credit card readers are also evolving. Apple has developed iPhones that can be turned into credit card readers. This is particularly attractive for sales outside of typical storefronts, such as carnivals, farmer's markets and street fairs.

In Japan, it's common to wave a cell phone over a card reader, which is so much faster than swiping a card through a terminal. At some point, this technology will gain popularity in the United States, but it may be a few years before the phone manufacturers, financial institutions, card terminal and card reader manufacturers and phone carriers can agree on who gets what. The showstopper in America is that the person who uses a cell phone is a customer of both the carrier and the bank, and that impacts how companies are compensated for the transaction.

This arrangement offers an advantage to the card companies. They can save money, because they don't need to send out cards to customers via the mail. Mailing credit cards always presents a security risk, because cards can be stolen from mailboxes or get lost during transit.

For retailers, verifying a transaction over a cell phone will be faster, and that's an added benefit for the customer. Account information can be placed within the telephone or card, and the added advantage is that you do not need to call over a network to send data. Customers may like having to only carry a cell phone with them instead of a wallet full of credit cards. Of course, cell phone transactions must be PIN-protected; otherwise, there is a real risk of fraud if a phone gets stolen. Yet, even with this risk, the ease of using a cell phone for payment and the speed in the checkout process may encourage more people to use their cell phones for spending.[85]

What is it going to take to have Americans paying with their cell phones in a few years? "[All] players will have to work together to define standards, determine revenue-sharing, expand the network of electronic readers, and think through the other parts of … this 2,000-piece puzzle,"[86] said Leslie Berlin, writing in the *New York Times.*

Contactless technology — in the form of stickers, prepaid devices, cameras, and phones — will become increasingly popular, because it offers quick and convenient services. "Contactless technology is a great evolutionary leap in payments," said Tom Allinger, Visa head of Global Product Innovation and Development. "It makes the payment into a chip that can sit anywhere."[87]

Analysts predict that high growth areas in Canada will include "mobile contactless payment, cross-border debits, and alternative methods for paying online."[88] According to one analyst report, contactless payments will be the high-test growth for Canada during this period, and cross-border payments will grow by 70% for the next five years.

---

85   Leslie Berlin, "Cell Phone Credit Cards Heading to U.S.," *San Jose Mercury News*, February 2, 2009, p. 4C.
86   Leslie Berlin, "Cellphones as Credit Cards? Americans Must Wait," *New York Times*, January 25, 2009.
87   "The Future Shape of Payments Is Anything but Flat," *American Banker*, February 24, 2009.
88   "Canadian Payments Market Geared for Growth Despite Troubled Economy," www.contactlessnews.com, January 21, 2009.

Current Trends

Banks are struggling to survive, and some public officials have even debated whether storied institutions such as Citigroup and Bank of America should be nationalized. The pressures on banks are trickling down to include their interactions with the credit card companies. Some changes taking place today will implement needed reforms, and others will help facilitate innovation. The following are some of the ways consumers and the payment card industry are being affected by these conditions.

Cardholder fees have increased. Issuers rely upon finance charge income for their profits. Late payment fees from some cards have been as high as 30%. For many years, the credit companies would go to great lengths to hang onto a loyal customer. Now, even long-term, loyal customers are subject to very high fees if they pay their credit card bills late.

A decrease in consumer spending is leading to consolidation in the credit card industry. To put this short-term spending decline in perspective, the housing slump and rising unemployment have impacted the earnings of the leading banks. In January 2009, credit card spending was down from the previous year by 17% for Bank of America, 15% for Citibank, and 8% for JP Morgan Chase. The 90-day delinquency rate has also become more frequent. Analysts are expecting more mergers in the credit card industry. Five card issuers dominate 80% of the business. This consolidation should offer greater opportunities for Visa and MasterCard.

Interest rates have risen. In 2009, Citibank customers were outraged when they received notices in the mail that the interest rates for revolving debt could accelerate. The notes came with a standard statement that said cardholders could opt out if they did not agree with the rates, but their credit cards would be discontinued in a short period of time. Although, for many years, consumers were inundated with mail offering attractive credit card rates, now they're more likely to get mail that informs them of rate hikes.

People are putting smaller, but more frequent, transactions on their cards. Issuers will be looking for ways to make money by increasing the opportunities for people to use credit or debit cards instead of cash. At merchants such as Starbucks, bold signs encourage people to use their cards for payment. Small transactions, under $25 — or at some merchants, under $50 — do not require signatures. Eliminating that requirement helps to accelerate processing and transaction time at the counter. This approach works well for convenience stores, movie theatres, fast food establishments, pharmacies, and other places where speedy transactions are especially important to business.

In tough times, businesses have to increase their focus on efficiency to survive. The Starbucks transaction model is very focused on efficiency. If customers aren't able to move quickly through the line, they will be much more likely to go elsewhere, especially during busy times. For small purchases, having an instant way to transact business eliminates the need to swipe and sign cards or to make change.

Qualifying for credit is increasingly difficult. Just as it's getting more difficult for people to qualify for loans, consumers will need to be more credit worthy before being offered a credit card, at least in the United States. "Lenders had become too reliant on consumers' FICO credit scores in the past. Now they are looking more at depreciation as a pattern to determine consumer behavior. Credit has been reduced for people in zip codes where homes are more likely to depreciate. Unfortunately, this makes it more difficult for the good consumer to get credit," said Meredith Whitney, in *The Wall Street Journal.*[89] Outreach programs designed to expand the use of credit and debit transactions in developing countries will help offset the economic impact of reduced spending in the United States.

Rewards programs are becoming more creative, and more discounts are being offered. Already, Visa has developed some unique ways of reaching its audience, such as offering discounts to high-end customers, if they

---

[89]   Meredith Whitney, "Credit Cards Are the Next Credit Crunch," *Wall Street Journal*, March 11, 2009.

submit information about their shopping habits. Visa is finding that customers are willing to give up personal information in exchange for money-saving coupons from their favorite retailers. Privacy is for sale, after all.

Credit card companies are offering better deals to businesses. Discover already offers businesses fee-free purchase checks that can be used at places that normally do not accept credit cards. American Express offers businesses discounts for early payments and provides options to pay 10% of the outstanding debt to extend the payback a couple of more months, interest-free.[90] Advanta introduced a card with no interest for 90 days on all business purchases, which is an advantage over the typical monthly float period from most cards. Some business credit cards offer lower interest rates than consumer cards offer. These business cards are available to customers who qualify based on their good business track record and personal credit history.

Balance transfer opportunities are harder to get. The typical zero-percent-interest-for-a-year offer has been used to encourage cardholders to transfer money from one lender to another. These offers are now made based on high credit scores. Most issuers are charging a balance transfer fee of 3% of the amount of the transaction, whereas many balance transfers used to be offered at no cost.

Financial institutions are more willing to work out payment plans for debt. They've already lost so much on bad home loans. They are more inclined than ever to forgive interest rates and other late payments in exchange for working out a viable payment plan.

The market for prepaid cards is wide open, particularly to the unbanked and under banked population. "…More than 35 million people in the United States do not have bank accounts; many instead use retail financial services such as check cashers"[91] This presents a great potential for the prepaid cards market.

90   Justin McHenry, "Trends in Business Credit Cards for 2009," www.smallbiztrends.com, January 17, 2009.
91   Analisa Nazareno, "Prepaid, Reloadable Payment Cards for Immigrants Roll Out," www.creditcards.com, September 19, 2008.

Chapter 6 - Simplify the Tax Code and Accelerate the Economy — One Transaction at a Time

While most people are understandably concerned about the impact of credit card regulations and controlling escalating interest rates, they also are often frustrated with the current tax system. It is complex, time-consuming, and costly. In fact, Americans spend more than 6.6 billion hours a year on tax returns.[92] There is so much paperwork that, in 2006 for example, the IRS published 582 *different* tax forms. The complexity in complying with the IRS Code costs Americans more than $100 billion a year. One Harvard economist predicted that implementing tax reform would boost the national wealth by nearly $5 trillion.[93]

Five trillion dollars! Imagine how all the work related to taxes would be simplified if you could reform income tax for consumers and businesses, and use electronic transactions to make that happen. Consider the impact it would have on our economy.

Electronic Transactions and the Flat Tax

This chapter explores a hypothetical concept. Let's assume the government decides to increase taxes using a flat tax approach, but wants to find a way to make it more acceptable by having people pay the taxes each time they make a transaction. This approach could not only reduce the costs of complying with IRS regulations, but it would also dramatically improve the cash flow of the government. Tax reform combined with a transaction-based economy could encourage business investment and ultimately reduce the amount of taxes overall.

Here's how this hypothetical approach could work. First, consider what might happen if money could be collected each time a transaction takes place. By leveraging electronic transactions to pay certain taxes, which

92   Michael C. Burgess, M.D., "H.R. 1040, *Freedom Flat Tax Act*," February 13, 2007.
93   Daniel J. Mitchell, Ph.D., "A Brief Guide to the Flat Tax," backgrounder Heritage Foundation, July 7, 2005.

would take a percentage of each transaction, the whole income tax collection process would be so much less painful. It would involve doing away with cash as we know it and replacing cash with electronic payments. This concept related to income taxes is not as difficult to implement or as farfetched as you might imagine. It also promotes a greener economy because everything is done electronically.

A popular way to address tax code issues first requires converting from the current system to one based on a flat tax for both individuals and businesses. There have been numerous proposals for a flat tax. For example, one proposal, *Congressional Bill, H.R. 1040*, the Freedom Flat Tax, would amend the IRS Tax Code of 1986 to provide a flat tax alternative. The bill would introduce a tax rate that starts at 19% for both individuals and businesses and declines to 17% after the first two years. This proposal repeals the estate, gift, and generation-skipping transfer taxes, and was recently reintroduced in February 2009.[94]

While some people think that a flat tax would discriminate against low-income families, the proposal makes provisions for lower-wage earners. A family of four would not be subject to the flat tax until they had an income of $36,600.[95]

A related plan, *The Flat Tax Act of 2009,* was introduced by Senator Arlen Specter on March 31, 2009. Specter's plan would eliminate the 17,000 pages of IRS code by implementing a flat tax of 20% for all individuals and businesses. Instead of taking an average of 34 hours to complete a 1040 form, the form for Specter's proposal would take about 15 minutes. It would simplify the payment of taxes, remove a lot of regulations, and reward savings and investments. This plan would eliminate taxes on estates, dividends, and capital gains, but maintain deductions for mortgage interest and charitable contributions.

The flat tax is a concept that has been debated since 1985, and can be fair, efficient, and effective. Instead of filling out pages upon pages of IRS

---

94   H.R. 1040 *Freedom Flat Tax Act,* introduced February 12, 2009.
95   H.R. 1040 *Freedom Flat Tax Act,* introduced February 12, 2009.

forms every year, individuals and businesses would simply complete 10 lines in a postcard-size form to determine their taxes. One form would cover individuals, and the other would cover businesses. It would be so simple to complete that you could determine your own taxes in about five minutes and never have to worry about keeping records for personal tax deductions, other than your income statements.

"At the business level, the tax would apply to the difference between sales of goods and services on the one hand, and the sum of wages, pension contributions, material costs and capital investments on the other,"[96] according to William Gale of the Brookings Institution.

Easy Money from Transactions

Electronic payments — such as those done via credit cards, debit cards, and prepaid cards — would help make this conversion to a business flat tax a reality.[97] They would be the means for collecting a portion of the flat tax. For example, whenever a customer purchased a certain product and paid for it electronically, a portion of the cost of the product would immediately be deducted from the seller's account at the point of the sale, and then reported to federal agencies, the banks, and the seller. This could be referred to as an eTax.

Putting It All Together with the eTax

Let's assume the eTax is 5%. So, if a computer company sells a PC to a retailer referred to as Company X for $1,000, the eTax on the PC is 5%, or $50. When the transaction is complete, $50 is electronically deducted at the point of sale from the specified computer company's account. The money immediately goes to the federal government and is tracked in numerous places. The seller (in this example, it's the computer company),

96   William Gale, "Business Taxes and the Flat Tax," Brookings Institution, www.brookings.edu, March 7, 2009.
97   William Gale, "Business Taxes and the Flat Tax," Brookings Institution, www.brookings.edu, March 27, 2009.

the government, and related parties get a monthly electronic report show-ing the taxes paid. That report is then used when the computer company reports its taxes at the end of the year.

So, when Company X sells that computer to the retail customer for $1,500, the same process applies to Company X. As a result, Company X pays a 5% tax ($75, in this example) to the government at the point of sale.

In essence, the computer company and Company X pay a portion of their taxes as transactions take place. The government benefits by getting a portion of taxes in advance, and the computer company and Company X do well because, ultimately, the flat tax rate results in lower taxes for their companies. They can immediately expense business capital equipment, and capital gains taxes are eliminated.

While the 5% rate deducted at the point of sale might not offset all of the taxes paid by the business, it would make a difference in what the compa-nies had to pay once a year and could be deducted from their flat taxes. The extra 5% taken by the government as each transaction occurs would dramatically improve the government's cash flow. The government would receive taxes at the time items are being sold, instead of waiting until each April 15 to receive payment.

What would it take to get companies to agree to do a pay-as-you-go form of taxation? The incentive would have to be a lower tax rate than they are currently paying, along with a simplified approach to accounting and more favorable expensing options. The downside, of course, would be if they were enticed to support such a program with a lower rate, and then, through the years, the eTax and the flat tax were to increase.

The Potential of the Flat Tax

A flat-tax proposal for businesses, along with the same type of approach for individuals, has great potential. It's simple and it supports business. When firms can deduct all of their capital expenses in the year they are made, that encourages investment and spurs the economy. With a flat tax, companies could make choices based on economic reasons, rather than making decisions just because of tax considerations.[98]

Many countries have applied a flat-tax approach, and they are benefitting from its results, particularly in the Baltic regions. The Heritage Foundation suggests, "In a global economy, it is increasingly easy for jobs and capital to escape high-tax nations and migrate to low-tax nations ... [A] flat tax will make America a magnet for investment and job creation."[99] Making these taxes transaction-based would only accelerate such growth.

The battle over the flat tax continues, and it may be years before this type of reform ever happens, if at all. While this discussion is primarily focused on flat tax and businesses, a similar process for electronically deducting taxes at the point of purchase could be implemented for consumers. They could receive periodic reports from their credit/debit/bank companies on the status of their eTax payments, and this could offset their tax liability. The advantage to consumers for having these deductions disrupt their cash flow would be the promise of a simplified tax process with ultimately lower taxes. Again, this would improve the cash flow of the government, while helping to reduce the amount of money consumers would have to pay in taxes each April.

Americans are already benefitting from the ability to file their taxes electronically and even pay their taxes on credit cards. A very successful American businessman I know uses the rewards points he gets for putting his taxes on American Express each year to take annual trips to Europe. Consumers can also reap other rewards from this hybrid approach of electronically paying a portion of their income taxes and business taxes.

The sample form below covers the basic questions that people would need to answer for the flat tax proposal (see Figure 7). Businesses would have a similarly simple form under a flat tax (see Figure 8). If an eTax is created

---

98   William Gale, "Business Taxes and the Flat Tax," Brookings Institution, www.brookings.edu, March 27, 2009.
99   Daniel J. Mitchell, Ph.D., "A Brief Guide to the Flat Tax," backgrounder Heritage Foundation, July 7, 2005.

to take out a portion of taxes in advance, simply add the eTax to the "Tax Already Paid" column." The use of electronic payments will only continue to expand, and the flat tax proposal is another innovative way to leverage this powerful technology and dynamic industry.

Simplicity on a Postcard: Sample Flat Tax Forms

Form 1	Individual Wage Tax	2005
Your first name and initial (if joint return, also give spouse's name and initial) last name		Your social security number
Home address (number and street, including apartment number or rural route)		Spouse's social security number
City, town, or post office, state and ZIP code	Your occupation	
	Spouse's occupation	

1. Wages, salary and pensions	1	
2. Personal allowances		
(a) $20,000 for married filing jointly	2(a)	
(b) $10,000 for single	2(b)	
(c) $13,000 for single head of household	2(c)	
3. Number of dependents not including spouse	3	
4. Personal allowances for dependents (line 3 multiplied by $6,000)	4	
5. Total personal allowances (line 2 plus line 4)	5	
6. Taxable wages (line 1 less line 5, if positive: otherwise zero)	6	
7. Tax (17% of line 6)	7	
8. Tax already paid	8	
9. Tax due (line 7 less line 8, if positive)	9	
10. Refund due (line 8 less line 7, if positive)	10	

SOURCE: Based on Robert Hall and Alvin Rabushka, *The Flat Tax* (Stanford: The Hoover Institution Press 1995).

Figure 7

Simplicity on a Postcard: Sample Flat Tax Forms

Form 2	Business Tax	2005

Business name	Employer identification number

Street address	County

City, town, or post office, state and ZIP code	Principal product

1. Gross revenue from sales	1
2. Allowable costs	
(a) Purchases of goods, services and materials	2(a)
(b) Wages, salaries, and retirement benefits	2(b)
(c) Purchases of capital equipment and land	2(c)
3. Total allowable costs (sum of lines 2(a), 2(b), and 2(c))	3
4. Taxable income (line 1 less line 3)	4
5. Tax (17% of line 4)	5
6. Carry-forward from 2004	6
7. Interest carry-forward (6% of line 6)	7
8. Carry-forward into 2005 (line 6 plus line 7)	8
9. Tax due (line 5 less line 8, if positive)	9
10. Carry forward into 2006 (line 8 less line 5, if positive)	10

SOURCE: Based on Robert Hall and Alvin Rabushka, *The Flat Tax* (Stanford: The Hoover Institution Press 1995).

Figure 8

Note: If the eTax concept becomes a reality, you could easily add a line to cover eTaxes paid and subtract it from the Tax due amount.

Chapter 7 - How the Credit Wars Will Be Fought

The current economic crisis will not last forever. Even Federal Reserve Chairman Ben Bernanke has been saying in 2009 that he expects the recession to turn around in 2010. As of October 2009, in fact, some economists are already saying that the turnaround has begun. The key players in the industry will all remain standing, along with some of the newer members, who will play a key role in shaping the direction of the industry. This chapter explores what you could expect to see happen with these companies through 2015. It looks at the following areas:

- Leading companies – Visa, MasterCard, American Express, and Discover

- Banks

- New companies yet to come

- PayPal

- Google

- First Data

- Revolution Money

- Phone Companies

Visa — The Strongest

Visa will continue to lead and shape the industry. On February 4, 2009, a time when most major companies on the New York Stock Exchange reported declines, Visa reported that its fiscal first-quarter profit rose by 35%, attributing this to the shift from cash to electronic payments. Payment transactions increased 13% that quarter.

By May 6, 2009, when MasterCard's first-quarter profit fell 18% due to what *The Wall Street Journal* called "swings in foreign exchange, lower investment income, and a slowdown in card spending,"[100] Visa managed to post strong Q2 2009 earnings. Visa's net income increased 13% over the prior year. Although revenue was slightly down in the United States, the volume grew in other regions. Keep in mind that no matter what happens with the economy, Visa still makes money from each debit and credit card transaction. Visa also projected a 20% rise in profit for its full 2009 fiscal year.[101]

Visa's biggest competitor, MasterCard, has lower projections for its own business in 2009. The company predicted its revenue growth will "fall short of its long-term objective of 12% to 15%."[102] In fact, KeyBanc Capital Markets had lowered its price targets on Visa and projected a 2009 revenue growth of just 5%.[103]

Even in the most difficult times, Visa prevailed because of its market share, brand, product value, and strategy. Visa represents 60% of all credit cards in the United States.[104] While banks are more selective regarding who can have a credit card, and they are lowering credit limits for some cardholders, there will always be a need for banks to offer credit cards as part of

---

100 Aparajita Saha-Bubna, "MasterCard's Profits Fall as Sales Dip," The Wall Street Journal, May 2-3, 2009, page B3.
101 "Visa Profit Rises 35 Percent Despite Recession," www.istockanalyst.com, February 4, 2009
102 Aparajita Saha-Bubna, "MasterCard's Profits Fall as Sales Dip," *The Wall Street Journal*, May 2-3, 2009, page B3.
103 "KeyBanc Lowers Estimates and Price Targets on Payment Processors MasterCard (MA) and Visa (V), www.streetinsider.com," January 15, 2009.
104 www.wikinvest.com/stock/Visa_(V).

their other services. When someone gets a home equity loan or opens a checking account, for example, that person often applies for a credit card as well.

Since Visa has such a huge market share, it looks like the company will remain strong. Worldwide, approximately 1.6 billion cards carry the Visa brand, compared to 900 million cards from MasterCard, and 90 million from American Express. In 2007, Visa handled 50 billion transactions.[105]

Visa generates revenues primarily from fees for card services, data processing, and international transactions. There are three parts to Visa's business: transaction processing, product platforms, and payments management. Transaction processing includes routing payment and related information, which is for authorizing, clearing, and settling transactions. The cards, with the Visa logo on them, are the product platforms. The payments network management includes promotions for Visa.[106]

"Everywhere you want to be," has been Visa's slogan for many years and is especially relevant as Visa reaches out to foreign markets for new opportunities. In fact, Visa's global network, VisaNet, reaches 170 countries and territories. The company's strategy is focused on growth. UBS analyst Adam Frisch and other Wall Street analysts count on "Visa's fee revenues growing 12% and earnings 20% annually well into the future."[107] Visa is moving into emerging markets, often areas where the concept of credit is truly foreign. The Dominican Republic, for example, uses Visa cards to provide aid to the poor. India represents a new opportunity for payments via mobile phones.

To support this growth, Visa will continue to innovate. The company will expand prepaid cards and store card operations. Visa is testing near-field communication (NCF) chips, for use in phones. Cell phones will be used

---

105 Stephanie Fitch, "Visa, the Future of Money," *Forbes Magazine*, October 27, 2008, page 86.
106 Bill Simpson, "Visa's IPO Fueled by International Growth and U.S. Shift to Electronic Payments," www.seekingalpha.com, March 20, 2008.
107 Stephanie Fitch, "Visa, the Future of Money," *Forbes Magazine*, October 27, 2008, page 88.

as credit cards, especially in emerging markets. The company is offering a range of new security processes, such as notifying card users when their teenagers' cards are used, which will help alert parents of any potentially crooked transaction.

In addition, Visa is offering a new card for upscale clients, Visa Signature, which provides special promotions, ticket presales, and access to VIP areas at events. Visa is also pursuing business customers, which had primarily been the stronghold of American Express.

About 85% of all e-commerce purchases in the United States are made on a major global payment network such as Visa. Globally, e-commerce represents more than $335 billion annually.[108] As e-commerce continues to grow, Visa will capture a dominant portion of this business. It captured nearly half of the $164 billion U.S. online spending in 2007.[109]

As Americans become increasingly dependent on e-commerce, Visa and other payment leaders will have greater opportunities. Of course, PayPal, Google (through Google Checkout), and other companies are expected to increase their market share, but the sheer number of Visa cardholders is a good indication that Visa will take the lead in this growth.

Each year, e-commerce transactions continue to grow. Even people who prefer to shop at local malls have discovered the advantages and convenience of purchasing items online, especially as some of the brick-and-mortar businesses fade with the rise of e-commerce sites. For example, local bookstore sales are declining, while online book sales are increasing.

Visa announced several innovations at its Future of Money Event in 2008. These include:

---

108 www.visa.com,
109 Stephanie Fitch, "Visa, the Future of Money," *Forbes Magazine*, October 27, 2008, page 94.

- Visa Applications for Android – Provides a suite of mobile applications that will allow cardholders to receive notifications on mobile devices about transaction activity, obtain offers, and use Google maps to find merchants that will redeem Visa or find ATMs.

- Visa-Nokia Collaboration on Next-Generation Handset – Provides payment and related services on near field communications (NFC) handsets for remote payments, money transfers, alerts, and other functions.

- Mobile Money Transfer Pilot Program – Allows people to use PDAs and mobile phones to securely send money to another Visa cardholder.

- Expansion of Prepaid Services – Provides a convenient way to add funds to reloadable prepaid cards, reaching financially underserved consumers.

Visa recently introduced the "Black Card," which targets high-end customers. For an annual fee of $495, the cardholder has a rewards program that offers a variety of options for points, such as hotels, airlines, and cars. It also offers limited access to lounges at airports, no annual minimum spending, luxury gifts, and concierge services for cardholders' special needs. The card is targeted to reach the top 1% income level. It offers a broad array of concierge services, such as golf reservations, sports entertainment tickets, business services, travel assistance, and specialty shopping for unique items.

Visa has experienced phenomenal growth, continues to innovate, and will maintain its leadership role in the industry. Even though the *Credit Cardholders' Bill of Rights* may impact earnings on fees in the future, the company is well positioned for growth because of the new markets, inno-vative uses of technology, expansion of e-commerce, and niche products reaching targeted groups. Visa has a strong platform, maintains good relationships with partners, and delivers a product that provides value and convenience.

MasterCard – A Winner

MasterCard, like Visa, has a big advantage in these tough economic times. Both companies get paid whenever transactions take place on their cards, but they don't have the same worry about the risks of consumers defaulting — risks that are faced by issuers, such as Citigroup or Bank of America. American Express is both an issuer and a payment company, so it has to deal with the impact of defaults.

MasterCard will not only survive, it will continue to prosper in the long term because of its innovative programs and extensive platform for credit. In 1991, the company launched Maestro®, the world's first online point-of-sale debit network. It launched the "priceless" advertising campaign in 1997, a powerful campaign with a slogan that has become a household term. The campaign appeared in 46 languages, indicating the reach of the company throughout the world. Then, in 2006, MasterCard began trading on the New York Stock Exchange as part of its new corporate governance and ownership structure.[110]

MasterCard faces growing competition from electronic funds transfer (EFT) or PIN-based networks. "This system may be perceived as more secure, since each transaction must be authorized at the time of sale, and are sometimes less expensive {to the merchant} because a flat fee is {usually} charged for each transaction. Visa currently holds a dominant market share in this sector,"[111] according to wikinvest.com.

---

110 www.mastercard.com
111 www.wikinvest.com/stock/Mastercard_(MA)

Like Visa, MasterCard grew revenue in 2008, a year when most companies saw their stocks battered and their revenues fall dramatically. Growth rose 10.7% worldwide in 2008, and processed transactions grew 11.8% to almost 21 billion. By December 2008, the company had issued 981 million cards, an increase of 7.6% over the previous year. End-of-year revenue reached a peak of $4,067 million and net income topped at $1,086 million.[112]

How is it that, despite reduced consumer spending, MasterCard still managed to be so profitable in 2008? The company was able to lower costs by cutting travel expenses, capping hiring, and reducing advertising.[113] As with Visa, cash is MasterCard's main competitor, so increased electronic payments in emerging markets represent a growth activity, even during challenging economic times.

MasterCard also has a strong reputation that will carry it through a period of economic turmoil. In 2009, MasterCard Worldwide was recognized as the "Best Corporate Cards and Expense Services provider" by *Global Finance* magazine for the third year in a row, because of its innovative solutions for providing value and efficiency in commercial payments.[114]

MasterCard considers itself a franchisor, processor, and advisor. As a franchisor, the company markets a vast portfolio of brands and products that are accepted at more than 28 million locations worldwide. As a processor, it provides global transaction processing across the MasterCard Worldwide Network. MasterCard also serves as an advisor, by providing insight and solutions to enhance the payment process.

MasterCard's PayPass product, which recently surpassed the issuance of 40 million cards or devices, is an innovative product designed to make

---

112   Ibid.
113   Jan Lagorio, "Visa, MasterCard Cuts Costs Amid Slowing Revenue Growth," *International Business Times,* February 2009.
114   Naya Larsson, "MasterCard Worldwide Remains Top Corporate Card and Expense Services Provider According to Global Finance Magazine," www. mastercard.com, March 12, 2009.

purchasing transactions easier and faster. Here's what the company's website says about PayPass:

> PayPass is a 'contactless' payment feature that provides con-
> sumers with a fast and convenient alternative to cash for their
> everyday small purchases, and can be added to a MasterCard or
> Maestro account.
>
> Consumers simply tap their PayPass-enabled card or device
> (such as a payment tag) on a specially equipped merchant termi-
> nal, eliminating the need to fumble for cash and coins, hand their
> card over to a clerk, or swipe the card. Signatures are not required
> for PayPass purchases under $25, further speeding customers'
> transactions.[115]

The PayPass device represents a major opportunity for the company to increase revenue through quick, easy, transactions that customers' desire. It also helps encourage people to use cards for services for which they would normally pay cash, such as fast food purchases. PayPass is also being used globally with rapid transit systems.

Like Visa, MasterCard provides concierge services, premium offers, unique travel benefits, and other comprehensive services with its credit products.

In Chapter 2, this book discussed the importance of educating students about credit. On its website, MasterCard.com, MasterCard offers a variety of credit-related educational materials, which are focused on college students. The site includes materials about managing budgets, understanding credit reports, dealing with debt, and so on. It also includes a "Step-by-Step Guide for Your New Financial Life." This service shows corporate responsibility and will ultimately help raise awareness of potential credit pitfalls, while helping students become responsible consumers.

---

115   Erica Harvill, "MasterCard PayPass Adds Ease and Simplicity to More Transactions Worldwide Surpassing 50-Million Issued Milestone, " www. mastercard.com, March 19, 2009.

On an international front, MasterCard continues to reach out to growing markets. "In the fourth quarter we saw ... U.S. gross dollar volumes and transactions for credit declined while debt still grew in the mid-single digits. We continued to experience volume and transaction growth in Europe in the mid-to-high single-digit range."[116] The company's South Asia, Middle East, Africa region is also still experiencing significant growth and that none of the major countries in that region has dipped into a decline. "Brazil continues to be a bright spot with solid, double-digit growth. In addition, our South Asia, Middle East, Africa region is also still experiencing significant growth,"[117]

MasterCard also reported strong growth across Latin America and the Caribbean, with volume up 11.4% on a local currency basis over the same period in 2007. In addition, the number of MasterCard-branded cards increased 16.6% in that region at the end of the fourth quarter of 2008.[118]

MasterCard will prevail due to its strategy of offering innovative products — such as PayPass, its cost cutting, increased use of electronic payments, and its ability to reach untapped markets.

American Express – Still in the Game

American Express (Amex) has been around in some form or another for more than 150 years. While MasterCard and Visa managed to prosper in 2009 despite the economic downturn, Amex's earnings plunged 79% in the quarter ending December 31, 2008 from what they were the same period the year before. Spending by cardholders fell 10% to 160.5 billion, and consumers and businesses struggled to keep up with their payments. Kenneth Chenault, chairman and chief executive officer of American

---

116   Weil, Judy, "MasterCard on Brazil and Europe," www.seekingalpha.com, February 9, 2009.
117   Judy Weil, "MasterCard on Brazil and Europe," www.seekingalpha.com, February 9, 2009.
118   Marcus Molina and Janet Rivera-Hernandez, "MasterCard Reports Strong Growth Across the Latin America and Caribbean Region in 2008," www. mastercard.com, February 11, 2009.

Express, warned of weaker spending, delinquencies, and uncollectable balances in 2009.[119]

Amex has considerable exposure to the domestic consumer markets, which can increase its vulnerability. It is also the leading issuer of commercial credit cards in the United States.

American Express accepted $3.39 billion in funding in late 2008 from the U.S. government as part of the Troubled Asset Relief Program (TARP) in exchange for preferred stock and warrants. This has helped to keep the company afloat. (Amex repaid the debt in June 2009). The company also announced in October 2008 that it would cut 10% of its workforce, which would generate a cost benefit of $1.8 billion.

Amex has come under criticism for using questionable criteria to reduce cardholders' credit lines. The company examined cardholders' spending patterns, and then searched for similarities to other cardholders who have had problems paying credit card bills. After complaints from cardholders, Amex decided to stop this practice.[120]

Amex is taking a variety of actions that may help compensate for some of the company's problems. For example, in March 2009, it reported $3.4 billion in new business signed for American Express Business Travel and a global retention rate of 98%. Global customer wins increased 172% year-over-year, and the North America middle-market segment grew 63% over the prior year.[121] This growth may have been the result of new programs, such as free travel program assessments, launched in 2008.

The company launched a practical rewards program in 2009 that lets cardholders earn double points on gasoline and grocery purchases up to $1,000 of spending each month. This program is a good option for dif-

119 David Ellis, "American Express Earnings Plunge 79%," www.money.cnn.com, January 26, 2009.
120 Ron Lieber, "American Express Kept a (Very) Watchful Eye on Charges," www.nytimes.com, January 31, 2009.
121 "American Express Travel Delivered More Than $2B in Travel and expense Savings to Customers in 2008 through Recession-Proofing Strategies and Solutions," www.americanexpress.com, March 12, 2009.

ficult economic times, because cardholders receive rewards for using their card on items they have to purchase anyway. The company also offers additional benefits, including consumer protections, such as warranty extensions, emergency cash, and roadside assistance.[122]

Discover — Hanging On

Discover Financial Services, launched in 1986, has never been able to catch up with the big guys. The company, formerly owned by Morgan Stanley and spun off as a separate company in 2007, operates the Discover Card, which it claims was the pioneer in providing cash rewards. It also offers student and personal loans and savings products. The company's payments business includes the Discover Network, PULSE ATM/Debit networks for the United States, and Diners Club International, a global payments network. Discover was also one of the earliest card companies to offer 24/7 customer service and a secure account number for online shopping.[123]

In June 2008, Discover sought about $6 million in damages from Visa and MasterCard, claiming the two giants "violated antitrust law and harmed Discover's business by preventing their member banks from issuing credit cards for Discover's network," as Martha Graybow reported in *Reuters*.[124] In November 2007, Visa had agreed to settle a similar lawsuit with American Express for about $2.1 billion.

Lawyers for Discover claimed that Visa and MasterCard shut their company out of the industry as it grew fivefold in 15 years.[125] Visa and MasterCard, however, claim that Discover's problems are based on its business model. They say that Discover hasn't entered into significant partnerships since

122 "American Express Offers Double Membership Rewards® Points on Gas and Grocery Purchases for Consumer Charge Cardmembers," www.americanexpress.com, March 24, 2009.
123 "Our Business," www.discoverfinancial.com,
124 Martha Graybow, "Discover Seeks $6bln in Case Vs. Visa, MasterCard," www.reuters.com, June 9, 2008.
125 "Discover vs. Visa and MasterCard," www.ecommerce-journal.com, September 24, 2008.

2004, which is partly because banks earn less from Discover than from major networks. Discover charges merchants less to use the card, and that equates to lower revenues for the bank partners.[126]

The antitrust suit was settled in October 2008. Visa and MasterCard agreed to pay Discover up to $2.75 billion in exchange for the company's agreement to dismiss the lawsuit, and hence avoid the issues related to a lengthy trial.

Position Statements in Discover's Antitrust Suit

The following are position statements by the three credit card companies involved in the 2008 antitrust suit filed by Discover against Visa and MasterCard.

Discover: "This settlement will enable Discover to further strengthen its capital base in this challenging economy and also will support continued investment in growing our business, including broadening global acceptance, expanding network volume and growing our deposit franchise..."

MasterCard: "We believe Discover's lack of success resulted from decisions that created a business model that is not attractive to bank issuers. Nonetheless, we chose to settle this lawsuit to avoid the uncertainty and distraction of a lengthy jury trial. This result, which is in no way an admission of liability, is in the best interest of our shareholders, our customers and our company.... We will continue to focus on out-competing Discover in the marketplace, where real-world performance is what counts..."

Visa: "Resolving this longstanding case on reasonable terms is in the best interest of Visa and our clients, cardholders and shareholders.... Visa will continue to focus on providing the superior value and reliability that our clients and cardholders have come to depend on..."[127]

---

126 Ibid.
127 "Discover Reaches $2.75 Billion Settlement with Visa, MasterCard," www.paymentsnews.com, October 27, 2008.

Discover has introduced some innovative products that should help promote the company's growth. In October 2008, for the third consecutive year, it was named in the Information Week 500 list of the most innovative users of information technology. The company was recognized for success with creating technology behind the new Discover® Motiva[SM] Card and for testing the viability of mobile banking.[128]

Discover.com Mobile, which was launched in January 2009, provided a mobile version of the Discover Card website that allows consumers to manage their credit accounts from their mobile phone's browser. Other innovative Discover products help cardholders control their finances with personal financial tools. The Paydown Planner, for example, calculates such things as the monthly payment required to pay down a balance over a certain timeframe and shows the impact of a large payment on the consumer's monthly statement.[129] The Spend Analyzer graphically illustrates the categories in which the consumer spends each month and provides monthly and yearly spending averages.

Discover also contributed to a green economy by being the first company in the United States to release a biodegradable credit card. The card is more expensive to create, but it does send a positive environmental message to consumers.

PayPal — Growing Stronger

While the credit companies will remain the leaders in the long term, PayPal is a popular alternative and will continue to increase its market share. PayPal has more than 65 million active users. "As technology evolves and improves, there will be more competition coming into the market for payment facilitation and processing, especially in eCommerce and Mobile

---

128 "InformationWeek 500 Ranking Highlights Discover Motiva[SM] Card Launch and Mobile Banking Demonstration," www.discoverfinancial.com, October 1, 2008.
129 "Discover Card Goes Mobile," www.discoverfinancial.com, January 15, 2009.

commerce (mCommerce),"[130] said Ed Kim, a business analyst and former consultant for Merrill Lynch.

To illustrate this growth, in 2007, PayPal had $47.5 billion in total payment volume [TPV], out of a total of approximately $113 billion in TPV for eBay, a 33% increase over 2006.[131] In 2008, that volume increased $60 billion.[132]

The advantage of PayPal is that it provides a quick and secure way to bill customers and accept payments online. Even without having a website, a merchant can still collect payments by creating an email invoice, which the customer pays securely through PayPal website. PayPal solutions are easy for merchants and consumers to use, and they provide merchants with reports, shipping, and sales tax tools that automatically calculate and collect shipping and taxes.

PayPal is attractive to merchants because PayPal's transaction fees are relatively low — 1.9% to 2.9% plus $0.30 per transaction — and do not require monthly set up or cancellation fees. Transactions are also PCI-compliant.[133]

Think of PayPal as an aggregator that issues credit to merchants. If you buy an item using PayPal, the company takes your money, bills the credit companies and gives money to the merchant. The company earns its fees by providing safe, secure online transactions that prevent fraud by protecting users from collecting each others' personal identification. One advantage for the merchant is that PayPal cannot put them on the Terminated Merchant File (TMF) list (see Chapter 4), also called the "Match" list. Merchants on the TMF list have had a merchant account terminated by a bank, processor, Visa or MasterCard for such things as fraud or excessive chargebacks.

130 Kim, Ed, "Visa, MasterCard Risk Ramped Up Competition," http://seekingalpha.com/article/70174-visa-mastercard-risk-ramped-up-competition, March 28, 2008.
131 Ibid.
132 Erin McCune, "eBay Analyst Day: PayPal World Domination," www.paymentsviews.com, March 11, 2009.
133 www.paypal.com.

Because PayPal offers some unique services that are not available through MasterCard and Visa, those two industry giants inadvertently helped to create the need for PayPal.

PayPal recently released a tool that allows e-commerce shoppers to use its service nearly everywhere online, even if the merchant doesn't accept PayPal. Shoppers can use a temporary MasterCard number, which they download, and the money is taken from the shopper's PayPal credit card or bank account. The merchant never sees the original account information. This is appealing, because it protects against credit card fraud and identity theft.[134]

The smartphone payments market also represents a tremendous growth opportunity for PayPal, especially since there were more than 263 million mobile phone subscribers as of June 2008. For example, PayPal offers software that runs on Apple's iPhone and has a payment option for people who purchase from Apple. The iPhone applications store was launched in July 2008; by January 2009, consumers had downloaded more than 500 million applications through the store.[135]

A really huge opportunity is PayPal's arrangement to be the only payment service provider for the application store that develops BlackBerry handheld devices. Purchases will go through PayPal's system.[136]

Smartphones are essentially tiny portable computers with a variety of communication features, such as text, web surfing, and email. These functionalities leverage PayPal's advantage as a service for online transactions. Smartphones also tend to have a friendly interface that encourages frequent use, increasing the potential for consumers to use them for payment transactions.

134 Catherine Holahan, "Credit Cards Lose Ground to PayPal," www.articles. moneycentral.msn.com, December 12, 2007.
135 David Wolfe, "PayPal Responds to Rapid Evolution in Smartphones," www. americanbanker.com, January 20, 2009.
136 Ibid.

It may be a few years before there is a mass acceptance of consumers using smartphones to make payments, but this will happen. To understand the future of smartphone payment adoption, recall consumers' initial reluctance to use ATMs. Today, consumers not only embrace ATM technology, but they take it to the next level with online payments from their checking accounts.

According to a January 2009 survey by Jevelin Strategy and Research, 70% of U.S. retail payments will be made online by debit and credit cards by 2013. That's a decline from an expected 80% in 2009. Email payments, such as payments made by PayPal, are expected to rise slightly from 9% in 2009 to 10% in 2013. The big gainer will be prepaid cards, with 5% of the market in 2009 and 13% by 2013. Increased security is the key reason why online buyers are starting to move from credit and debit cards to payment services.[137] So, while the increase for PayPal in U.S. online retail volume is not very significant, it is still growing. Part of PayPal's strategy entails going after larger retailers. The company intends to nearly double its payment processing volume over the next three years in a variety of ways by increasing penetration on eBay, growing its share in merchant services, and expanding to markets outside of e-commerce, such as mobile payments, nonprofits, and social networks.

According to a report by Jupiter Research, PayPal use is far more popular than alternative payment methods from other companies. In one survey, 23% of online customers preferred PayPal, while 4% preferred BillMeLater, and 1% preferred Google Checkout. While security was the top reason for consumers to use alternative payment methods, the next driver was that they had no other payment options. For example, PayPal is the only method allowed for purchasing through eBay.[138]

PayPal provides a way for people who do not have their own custom terminals to process credit card payments. PayPal has made it possible for the average person to easily sell an item on eBay and be confident in

137 "How Do Online Buyers Pay?" www.emarketer.com, January 14, 2009.
138 Susan Kuchinskas, "PayPal's Grown-Up Growth Outlook," www.internetnews.com, March 6, 2008.

getting paid. Without this service, checks would have been too risky and time-consuming. PayPal was the answer for this market.

Google Checkout — Second Best

Google Checkout has been around since 2006. Google, of course, knows how to dominate a market and succeed. Yet, although Google plays a big role in payment processing, it lacks some of the strengths of PayPal.

Let's look at some of the comparisons between the two products. In terms of features, PayPal is more commonly selected because of its flexibility. PayPal accepts e-checks and makes direct deductions from bank accounts. An e-check is an electronic version of a paper check. Just a like standard check, the e-check has a number assigned to it for tracking purposes. Google, however only takes credit or debit cards.[139]

PayPal does business globally and also does currency exchanges. Google Checkout is focused on business in the U.S.

PayPal offers better customer service than Google. If you have a question about your account, you can call PayPal and speak to a live person. Have you ever tried to talk directly to an individual at Google or tried to call someone from the customer service department to solve a problem? Chances are you never reached a human, because Google is focused almost exclusively on providing interactive online service. While more and more companies are trying to eliminate the cost of responding to a call, and many people prefer web-based self-service, there is something to be said for having the opportunity to talk to a human being.

Some areas where Google shines, however, include Google's fraud protection policy, which offers 100% refund within 60 days on all items. PayPal won't offer that protection unless the purchase is over $50.[140] Yet,

---

139 "Google Checkout Vs. PayPal," www.bestshoppingcartreviews.com, April 17, 2009.
140 Ibid.

both services offer secure sockets layer (SSL) security for data protection that is comparable to that of banks.

Google does have an advantage with AdWords, which provides discounts on services, whose costs can be deducted from AdWord earnings. Also, if the product you wish to purchase appears on a search or in an ad, a shopping cart icon will appear. As a result, you can buy directly from the search page.

The cost of Google Checkout is an issue. The company recently announced it would change the basic price for Google Checkout of 2.2% plus 20 cents into a four-tier pricing plan. This increased the costs for most merchants, except for those who submit 100,000 or more Google Checkout transactions per month. The company also dropped its perk of $10 in free Checkout processing for every dollar a merchant spends on Google AdWords.[141] This increase could turn out to be a problem for Google, because lower processing fees were the key to Google capturing business from PayPal. As processing fees increase, Google could become less attractive to merchants than PayPal, although the ability to use an AdWords account may help reduce the cost of Google Checkout transaction fees.

Overall, Google and PayPal share a few things in common. Both make it easy to buy items over the Internet for a low price. As third-party processors, they shield customer and cardholder information from retailers.

One area of growth involves using Android as the new operating platform for a mobile phone. By the end of 2009, at least 18 phones based on the Android operating system will be available. In fact, Visa is developing an application for Android that lets mobile users pay for goods using their phone and receive information about their bank accounts. It can leverage location-based technology to determine where a person with a phone is located and then send them information about retailers nearby that accept Visa offers. "This is lucrative for Visa because they will be able to target

141 "Merchants Vent Their Ire Over Checkout Pricing on Google Site," www. digitaltransactions.net, March 25, 2009.

users with various offers based on previous purchases. Visa is looking to the long term. With over 3 billion mobile devices being used around the world at the moment ... the payment processing giant is looking to make sure they are not only part of, but leading this technology shift that is already happening. This is also good news for Google and Android as it is another selling point for the mobile platform in such a competitive market."[142]

First Data — The $26-Billion-Pound Gorilla

First Data is a global technology leader in information commerce that helps businesses, such as financial institutions and merchants, process customer transactions and understand the information related to them. Think of this company as the gorilla that does the business' back-end work. It does the heavy lifting. The company is doing well, even in this challenging economic environment, and reported merchant services revenue of $1.1 billion, up 18%, with full-year merchant services revenue of $4.1 billion, up 10%.[143]

With First Data, transactions move quickly and securely. The company is trying to capture a big chunk of the credit card business, and it may make a dent. In fact, First Data has the infrastructure to compete with the card associations and could be the next seriously competitive company in the card business.

In 1971, First Data Resources (FDR) incorporated to become a for-profit organization, providing processing services to MABA. FDR had 110 employees and $2 million in annual revenues.

The name First Data may not be familiar to people outside the credit industry, but you may be surprised that it includes another company known to nearly everyone, Western Union. By 1980, American Express

---

142 Patrick Martin, "Your Android Phone Will Replace Your Wallet," www. androidmobilephone.com, October 9, 2008.
143 "Acquiring Gains at First Data Despite a Tough Merchant Environment," www.digitaltransactions.net/newsstory.cfm?newsid=2125, March 26, 2009.

Information Services Corporation had bought most of First Data. First Data was subsequently spun off from American Express and went public in 1992. In 1995, the company merged with First Financial Management Corp. (FFMC). Western Union became part of First Data as a result of that merger.[144]

By 1998, First Data was providing services worldwide. From 2001 through 2005, First Data acquired companies and added the STAR Network and PIN-based debit acceptance at more than 2.1 million ATM and retail locations. In 2006, the company spun off Western Union.

In 2007, First Data was acquired by an affiliate of Kohlberg Kravis Roberts & Co. (KKR) for $29 billion.[145] It was a huge price tag, but First Data had been referred to as a "free-flow cash machine" that had excellent growth prospects and great international business potential. A growth opportunity was in the dynamic currency conversion area, where a bill on a foreign card would be immediately converted to the customer's local currency. Without this capability, a customer might not know what the Euro converted to until he got his statement. This capability provides the issuer, the acquiring bank, and the merchant with revenue. First Data had the infrastructure and relationships in place to take advantage of dynamic currency conversion.[146]

In 2008, the company released a successful contactless payment device, the Go-Tag, which is a sticker that also functions as a contactless prepaid debit card. The tags are about the size of a quarter, and people stick them on something they're used to carrying around, such as a cell phone, keys, or wallet.

"Worldwide, there are about 3 billion mobile phones in use, and hundreds of millions more are added each quarter, Mr. McCarthy [president of mobile solutions for First Data Corp] said. In contrast, there are about 1.4 billion payment cards that are actively used," according to Will Wade, writing

---

144 www.firstdata.com/abouthistory.htm.
145 Ibid.
146 Georges Yared, "KKR and First Data: Why a Payment Processor Is Worth $26 Billion, www.bloggingstocks.com, April 1, 2007.

for Mobile Banker. "Those numbers make the mobile phone extremely attractive to payments executives. Turning mobile phones into payments devices 'can triple the size of the electronic payments opportunity,' Mr. McCarthy said."[147]

First Data is also making headway into the prepaid card market. In March 2009, First Data and Urban Trust Bank announced that they will offer the Money Network payroll card and related prepaid products. The Money Network card offers employers a cheaper alternative to issuing paper paychecks.[148] This offer is reaching the "under-banked" market, which is expected to show significant growth over the next few years and will compete head-to-head with the likes of Visa, MasterCard, and others.

Another opportunity for First Data is in Asia, where First Data allows small businesses to accept card payments in the store, online or by phone, and in local and foreign currencies. "Card transactions skyrocketed over the year – and our mission is to ensure that Asian businesses are equipped with the tools to accept the widest possible range of card payments,"[149] said Merchant Solutions CEO and Managing Director Sean Hesh in a First Data press release.

Revolution Money — The Newbie Innovator

Never underestimate the power of new, innovative companies that enter the market. Revolution Money, for example, is an alternative payment company that offers secure, easy, and flexible payment options for merchants and cardholders. It was founded in 2005 by Steve Case and designed to give consumers more options. It is known for its low-cost payment card network.

---

147   Will Wade, "The Future Shape of Payments Is Anything but Flat," *American Banker*, February 24, 2009
148   "Urban Trust Bank and First Data Partner to Offer Prepaid Cards," www.reuters.com, March 3, 2009.
149   "Merchant Solutions Enables Tens of Thousands of Asian Businesses to Accept Electronic Payments in First Year," www.firstdata.com, November 19, 2008.

Revolution Money is the largest PIN-based network in the U.S. and offers the first PIN-based credit card in the nation.[150] What makes Revolution Money particularly attractive is that it eliminates the interchange fees that merchants would have to pay to a traditional payment processor or bank for accepting cards. Revolution Money processes its Revolution Card transactions on an internet-based system, allowing them to reduce its processing fees and pass the savings on to Merchants. The company focuses on a variety of vertical markets, such as cable, electronics, groceries, and sports. It also provides an online peer-to-peer payment service where members can send and receive money from other members at no cost, with the exception of some fee-based ancillary benefits of receiving paper statements, making balance inquiries, and using their ATM debit card for withdrawals.[151]

Cards issued by Revolution Money do not contain customer names or other identifying information. As a result, the card is less likely to be subject to identity theft and fraud.

Mobile Phone Technology Leaders

Smartphones provide merchants with a tremendous potential for increasing sales, accelerating collections, and reducing bad credit debt. Treos, iPhones, and other smartphones enable users to purchase credit card processing applications, take credit card payments, and transmit them securely — all without magnetic stripe readers or other devices. Merchants have the flexibility to work anywhere without having to worry about extra landlines and processing equipment.

As these applications become standard on smart phones, they are increasingly appealing to business owners and companies. InnerFence, an iPhone application development company, has created an iPhone credit

---

150 "Chase Paymentech and Revolution Money Expand RevolutionCard Merchant Acceptance," www.reuters.com, March 30, 2009.

151 www.revolutionmoney.com

card terminal. Processing transactions with a smartphone is very much like doing online transactions, because they do not require magnetic stripe readers. Payments still need to be transmitted through secure payment processors before the sale goes through, however.

According to Shyam Krishanan, a technology analyst with Frost and Sullivan, "As the economy picks up and as people decide to spend more on high-end items, these applications will come into reckoning.... This would mean lesser usage of credit/debit cards as consumers view the phone as a one-stop tool for all of their requirements."[152]

Not to be outdone, Visa just introduced a cell phone payment system that uses a chip on the phone to communicate with a payment terminal. This version is being introduced in Malaysia and is based on a new near field communication (NFC) global standard for phones and telephones.

With this new technology, consumers can load multiple accounts onto one phone. In Malaysia, for example, consumers can use their cell phones to pay for parking and public transit fares. "Eventually, the system will allow credit and debit card accounts from multiple providers and payment brands. You need to run an application on your phone to switch the default account that is charged when you swipe the phone over a terminal,"[153] writes Amy E. Buttell on CreditCards.com.

Visa is also working with a large group to further its involvement in mobile, contactless technology. In September 2008, Credit Suisse, PostFinance, SIX Multipay, Swisscard, Swisscom, and Visa Europe tested contactless payment using NFC technology for payments made at the point of sale by mobile phones and credit cards. "The mobile phone solution is an innovative enhancement based on Visa payWave, the contactless payment method of Visa," [154] Credit Suisse said in a news release.

---

152 Amy E. Buttell, "Merchants Eye Mobile Phones to Transact Card Payments," www.creditcards.com, March 17, 2009.
153 Ibid.
154 "Contactless Payment by Mobile Phone — Technically Possible, Work Continues on Requirements for Commercial Use," press release, www.creditsuisse.com, April 29, 2009.

Results indicated that paying by mobile phone works from a technical point of view, is easy and convenient, and meets customer objectives. The Credit Suisse news release continued, "Using the test findings as a basis, a working group consisting of companies from the financial industry and a recently created working group consisting of mobile telecommunications companies are now collaborating with merchants to produce solutions that are viable for all the acceptance point partners. The aim is to define the exact requirements so that in future, clients can use their mobile telephone to make payments in as many places as possible."[155]

Phone Companies

While several phone companies are attempting to move into the payment card business, AT&T clearly plays an important role. In June 2008, the company joined with Citi to launch a new credit card for small business owners. In addition, the company offers mobility solutions for business.

As more transactions move online with mobile payments, Verizon, in particular, stands to benefit. Verizon currently provides merchant services with discounted processing rates for larger merchants. The company is targeting Verizon business customers and offering the ability to quickly and reliably process transactions. Verizon provides the rapid broadband connection and convenient credit card transaction processing from Chase. It also offers to eliminate the expense of a dedicated card terminal phone line, because wireless terminals are available.

Creating a New Currency

The credit card, debit card, and prepaid card have become a portable, personal cash machine for most people - much handier than paper bills and coins. In fact, the Treasury Department is thinking of getting rid of the penny and other coins, because it costs more to produce them than they are actually worth.

155 Ibid.

Cash is also inconvenient. It has to be counted, which takes more time than processing an electronic transaction. Think about it. How many times have you waited in line to pay for an item only to be frustrated by someone ahead of you who dumped out a huge amount of coins to pay for a purchase? Haven't you seen a sales clerk having to count each coin before completing a transaction, while people in line wait impatiently for their turn to purchase goods?

Money is inconvenient. It consumes valuable resources — paper, ink, and metals — and carrying a whole wad of bills or a handful of coins can weigh down your purse, pocket, or wallet. If the money gets stolen or lost, you're out of luck. If you pay by cash and lose the receipt, you might have problems returning an item. Yet, even if you lose a receipt for a credit or debit payment, you can always go on line and get a statement that lists that purchase.

So, non-cash payments are light, eco-friendly, more secure, and make it easier to return items if necessary. About the only advantage of using cash is that your purchases are anonymous, but most people are not concerned about keeping purchases anonymous. In fact, an advantage of card payments is that you can track your bills and know where your money has been spent. If you use your cards wisely, you can become more aware of what you consume and do more effective budgeting.

Previously viewed as perks for a select few, payment cards have become so much more than that. Credit and debit cards have become their own form of currency, as people increasingly favor using them over using cash. As the use of credit cards continues to rise, the United States is getting closer to becoming a "cashless" society. The time to authorize transactions is down to a few seconds, making it a speedy, efficient way to conduct business.

An even faster way to conduct business is with the new, emerging technologies that allow customers to simply wave a cell phone or other device near a terminal, making checkout even faster. In fact, in the future, cell phones will be so smart that a store will be able to identify you from your mobile device as you walk down the grocery aisle, and screens on the shelves nearby could start to "come alive" by talking to you and recommending purchases based on your recent spending habits.

Within the next 7 or 8 years, you might be able to do your shopping without even standing in line. "Whatever replaces the legacy point of sale terminal will undoubtedly be based on mobile technology and will use RFID [radio frequency identification], Internet protocol or both to complete transactions in the blink of an eye. Today if you used RFID in its purest form, you could walk into a store, load your cart, and walk out without talking to anybody because they would know who you are,"[156] suggests Jay MacDonald, writing for CreditCards.com.

The end of cash also contributes to the demise of checks. In 1999, checks accounted for 49% of consumer payments, with credit at 23%. By the end of 2009, the percentage of transactions done with checks is estimated to shrink to 18%, with credit and debit card transactions at 45%.[157] In fact, about 13% of Americans don't even have a checking account.

By 2012, two-thirds of all bills will be paid electronically. The benefits include quick payment verification, convenience, and lower cost to the consumer. No postage is needed. The main reason that people pay this way is to control the timing of the payment.[158] This convenience motivates people to use payment cards.

When we make the leap to an electronic society, it's important that back-up systems are in place so that we are not stymied when a natural disaster or crime causes electronic activity to come to a screeching halt. A case in point is a recent incident in the San Jose area, which occurred on April 9, 2009. Some fiber optic cables were cut due to sabotage, shutting down AT&T's phone, Internet, and wireless services to consumers. ATM ma-

156 Jay MacDonald, "What Will Credit Cards Look Like in 25, 50, or 100 Years?" www.creditcards.com, February 17, 2009.
157 Nick Connors, "Consumer Payment Trends Favor Plastic Payment Methods," www.creditunion.com, October 27, 2008.
158 Kathi Plymouth, and Jodi Martin, "Bill Payment Trends: Major Shifts in Consumer Behavior Require Comprehensive Planning," www.firstdata.com, 2009.

chines didn't work, and many local businesses couldn't accept payment cards since phone and cable lines were down. This had an impact on local businesses, as consumers left stores unable to complete their purchases. Although the problem was fixed and confined to a small region, imagine what would happen if such an incident were widespread.

This is just one, isolated instance that shouldn't have happened, and there are plans to prevent this from occurring again in San Jose and elsewhere. With the right controls to prevent fraud, natural disasters, and sabotage from impacting the ability to conduct business with payment cards and other electronic transactions, we are in a good position to maximize the benefits of being a cashless society.

At some point, we will transition fully beyond plastic. A future payment card, for example, may not even be a card at all. One designer came up with the concept of a way to make purchases more secure and to use technology as a personal shopping advisor. It would allow you to know your balance before you make a purchase, and this would help you avoid overspending. It can use RFID technology and approve purchases with a built-in biometric fingerprint reader.[159]

Payment-enabled phones probably won't become mainstream in the United States for another 10 years, but they will help take plastic to the next level. As David Greer opines on CreditCards.com, "With 2.7 billion cell phones in circulation globally, credit issuers have plenty of motives for turning the gadgets into contactless credit devices."[160]

---

159   Ido Genuth, "Future Credit Card," www.thefutureofthings.com, May 18, 2007.
160   David Geer, "Cell Phone: Contactless Credit Card + Yak, Wave, Buy," www.creditcards.com, January 11, 2008.

Chapter 8 - Conclusion — Continued Growth and Innovation

We can only look ahead at the new opportunities that the payment industry brings to businesses and people. Payment cards are the foundation of many small businesses in the United States. Without them, many companies would never get a chance to exist. Many students wouldn't be able to pay their college tuition. Health consumers might not have the money for emergencies, such as medical procedures. While consumers need to avoid spending beyond what they can pay back each month, they can learn to use credit responsibly.

As the banks cut back on the credit offered to consumers, people in the United States will move from overextending themselves and will seek a more responsible approach to consumerism. They will be more likely to pay bills off each month to avoid interest payments. They will also become more knowledgeable about credit.

In 2010, the new *Credit Cardholders' Bill of Rights* will ensure that details about rates are more clear and concise in credit card offers so that people understand how much interest they will have to pay. Because the bill sets limits on interest rates, cardholders will benefit.

The increasing use of credit will offer new opportunities to players in the industry. It's no surprise that more businesses will move online over the next few years. Nielsen discovered significant year-over-year (ending September '08) growth in online activities with an increase of 17% in time spent on e-commerce sites.[161]

As this transition happens, there will be more opportunities for payment cards over cash, because online transactions use cards or e-checks. Fraud prevention technology will help to make these transactions even safer than transactions in a retail environment where the credit card is swiped. This technology will identify potentially fraudulent or risky cards

---

161 "2009 U.S. Industry Outlook, 'When Times Get Tough, the Tough Go Back to Basics," Nielsen, December 2008.

and prevent fraudulent transactions from occurring. It will help consumers keep their cards secure and reduce the risk for merchants.

It's not unusual to see newspaper headlines focus on the problems of debt and credit in America. We're just beginning to climb out of a recession in which a number of banks have failed. Foreclosures are still happening all around us. Unemployment is still in the double digits. All of these negative factors could lead some people to believe that the credit card industry is battered. Yet that way of thinking is shortsighted. The leading companies that emerge from this recession, such as MasterCard and Visa, will become stronger, because they already have the required infrastructure in place, and they provide a much-needed service. Even if people cut back on spending, the leading credit companies and merchant processors will make considerable profits via the payment card business.

New markets will open. For example, it wasn't too long ago that doctors, lawyers, accountants, and other professional service groups didn't take credit cards. Now we are discovering that more businesses than ever before are accepting payment cards, because their customers demand it. As the costs to process checks increase, companies will step up their efforts to encourage the use of card payments. Some airlines, for example, are already doing this by refusing cash payments for purchases made on board.

On the other hand, small businesses, which have been so reliant upon credit cards, will suffer some setbacks due to impending legislation. They will face challenges similar to those of the consumer in the short term — smaller credit lines, fewer rewards, and fewer online offers. For example, Chase has (at least temporarily) eliminated most of its business credit cards from online channels.[162]

While the new *Credit Cardholders' Bill of Rights* does not offer small businesses the same type of protection that it offers consumers, advocacy groups are working to extend this protection to small businesses.

---

162 McHenry, Justin, "Trends in Business Credit Cards for 2009," www. smallbiztrends.com, January 17, 2009.

Business owners will continue to be offered better rates than consumers, if they have a good track record. They will also receive additional business discounts on products and services. In addition, the credit card industry will offer them greater opportunities to purchase relevant products. For example, Discover offers purchase checks that are used to buy goods from companies that don't take credit cards. American Express will provide 2% discounts for paying early, or allow small businesses to pay just 10% of their debt in order to extend the payment periods for two more months, interest free. Other cards increase the interest float on businesses to 90 days instead of the average float of 20–30 days.[163]

Other new market opportunities include expansion into many countries, and offering payment card and related services to the unbanked and under-banked customers. They also include targeted programs with perks for affluent customers and for businesses.

In this new, competitive world of payment cards and other electronic transactions, the companies with the infrastructure and services that people need will prevail. Visa will continue to grow because of its effective business model, unique offerings, and valuable service. It will be followed by MasterCard. PayPal will remain a player because of the niche payment capabilities and reduced risk it offers to customers. First Data may even have the muscle to become the next big card company.

What will happen to the merchant processors? While retail businesses continue to decrease, they will find new opportunities in the growing online business with companies that previously had not accepted payment cards.

As Americans scale back on their average card purchases, people in emerging countries will continue to increase their spending. While the average transaction dollar value will be less internationally, the sheer number of people using payment cards will increase the volume of transactions. The increase in transactions will contribute to growth in the payment industry.

163 Ibid.

The most innovative companies will make a difference. Consumers will embrace new ways of making payments — whether it's via a card, a bio-chip, or a smartphone. Whatever makes the transaction easier and faster will help the industry to grow. The easier it is to perform transactions, the more likely consumers will be to use payment cards. That's why mobile payments from cell phones are so attractive to consumers.

Credit card companies and banks will take measures to increase consumer awareness. Earlier in the book, I proposed that people complete a test before being able to receive and use a credit card. While that may seem restrictive now, it has many advantages and is a realistic proposal, especially considering the number of students who are issued cards and then leave college saddled with credit card debt.

Our culture may have tempted too many people to spend beyond their means for far too long. Heavy credit card use and subprime loans may have contributed to their personal financial losses we are seeing today. Yet the turmoil in our country's economic climate will make them better consumers. Maybe they will put fewer charges on their cards. Maybe they will pay off bills in full each month. Ideally, they will use credit more wisely and will raise the next generation to understand the impact — and the value — of a cashless society.

Glossary

Acquirer: An organization licensed as a member of Visa/MasterCard as an affiliated bank or bank/processor alliance. The acquirer may also set up new merchant accounts and process credit card transactions for merchants/businesses (acceptors).

Authorization: Verifies the credit card has sufficient funds available to cover a transaction.

Authorization Response: An electronic message from an issuing financial institution in response to an authorization request for a card transaction. The response may result in approving or declining the transaction or it could involve sending the response to a call center for more information.

AVS: Address Verification System. A service provided by payment card networks to verify the cardholder's address associated with the card being used. AVS is one tool used to mitigate fraud.

BIN (Bank Identification Number): The first six digits of a card number. It identifies the issuing bank and card network to which it belongs.

Card Issuing Bank or Financial Institution: A bank or other institution which offers branded payment cards to customers.

Cardholder: Individual who holds a payment card account. The account is used to purchase goods or services.

Card Network: The technology system which authorizes and captures credit card transactions.

Card Not Present: A transaction in which a card is not present, such as through a web site or over the telephone.

Card Security Code: The 3- or 4-digit code required for card-not-present transactions to minimize the risk of fraud. This code is commonly referred to as CVV or CV2 (Card Verification Value) or CID (Card Identification number). Visa, MasterCard, Diners Club, Discover, and JCB use a 3-digit code, while American Express uses a 4-digit code.

Chargeback: The return of funds to a cardholder, usually resulting from a consumer disputing a transaction or service from the merchant. The merchant's bank account is debited for the transaction after the sale has been settled.

CISP: Cardholder Information Security Program (Visa's data security program).

DISC: Data Security Guidelines (Discover's data security program).

Debit Card: Payment card that enables deductions directly from the card-holder's checking account.

Discount Fee: The percentage of sales charged to a merchant for each card transaction.

DSS: Data Security System (American Express's data security program).

EFT: Electronic Funds Transfer.

Independent Sales Organization (ISO): A company that represents a bank or bank/processor alliance. The ISO (if it is also an acquirer) can open merchant accounts and process card transactions. An ISO has control over the fees it charges to merchants as well.

Interchange: This involves an electronic exchange of financial and non-financial data related to sale and credit data between merchant acquirers and card issuers. This applies to MasterCard and Visa credit card transactions.

Interchange Fee: A fee paid by an acquirer to an issuer for MasterCard and Visa credit card transactions.

Offline Debit: Transactions using debit cards with a Visa or MasterCard logo. Funds transfer is usually delayed 2-3 days, the typical time for Visa/MasterCard settlement into a merchant's account.

Online Debit: Transactions requiring a PIN authorization. Funds transfer is typically reflected immediately on cardholder's bank account.

Payment Gateway: Technology which protects sensitive card information, such as credit card numbers, during card-not-present transactions. A payment gateway enables the transfer of information between a payment portal (such as a website or telephone service) and the acquiring bank.

PCI: Payment Card Industry.

PCI DSS: PCI Data Security Standard. The set of security standards set forth, by the PCI Security Standards Council, for the safety and protection of cardholder data. There are 12 main requirements which must be followed by merchants and payment service providers who process and store credit card data.

PCI SSC: PCI Security Standards Council. Founded by payment brands American Express, Discover, JCB International, MasterCard, and Visa. The council's purpose is to develop, manage, and ensure global data security measures.

Point Of Sale (POS): A location, such as retail store, where a cardholder is present to conduct a transaction and swipes a card.

POS Terminal: Equipment that collects, transits, and stores credit card transactions.

Processor: A company that routes card transaction authorization to Visa or MasterCard. The processor also manages the funds settlement to the merchant.

SDP: Site Data Protection (MasterCard's data security program).

Settlement: What occurs when the acquiring bank and the issuing bank exchange data or funds following a card payment transaction. Most commonly, settlement is referred to as the deposit into, or deduction from, the merchant's bank account.

Bibliography

AARP Outreach & Service," Credit Card Fraud," www.aarp.org, February 20, 2009.

"About Revolution Money," www.revolutionmoney.com.

"Acquiring Gains at First Data Despite a Tough Merchant Environment," www.digitaltransaction.net, March 26, 2009.

AmericanExpress.com

"American Express Company – DBRS Confirms American Express Company – Senior at A (high), Changes Trend to Negative," www.dbrs. com, January 28, 2009.

"American Express Offers Double Membership Rewards Points on Gas and Grocery purchase for Consumer Charge Card Members," press release, www.americanexpress.com, March 24, 2009.

"American Express Reveals Exclusive Card Member Experiences Under the Tents at Mercedes-Benz Fashion Week in New York," press release, www.americanexpress.com, January 22, 2009.

"American Express Company," www.nytimes.com, April 2, 2009.

"AT&T and Citi Launch New Credit Card for Small Business Owners," www.att.com, June 5, 2008.

"AT&T Rolls Out New Credit Card Rewards Program for Small Business," www.sanantono.bizjournals.com, June 5, 2008.

Barr, Alistair, "Credit Card Industry Faces Tough 2009," www.menafn.com, January 23, 2009.

Barrett, Amy and Quittner, Jeremy, "Cash Vanishes from Merchants' Accounts," www.businessweek.com, April 24, 2009.

Berlin, Leslie, "Cellphones as Credit Cards? Americans Must Wait," www.nytimes.com, January 25, 2009.

Berlin, Leslie, "Cell Phone Credit Cards Heading to U.S." *San Jose Mercury News*, February 2, 2009, 4C.

Borland, John, "Malware Swipes Millions of Credit Cards," www.technologyreview.com, January 22, 2009.

Boyd, Roddy, "Visa Can Credit Rivals for Rich IPO: Visa's Blockbuster IPO Owes Thanks to the Success of MasterCard's IPO and to the Credit Debacle Plaguing Rivals Like American Express and Discover Financial Services," *Fortune Magazine,* 2008, republished at www.cnnmoney.com, February 26, 2008.

Burgess, Michael G., H.R. 1040, *Freedom Flat Tax Act*, February 13, 2007.

Buttell, Amy E., "Merchants Eye Mobile Phones to Transact Card Payments, "www.creditcards.com, March 17, 2009.

"Canadian Card Payments Forecast 2009-13," www.buildaskill.com, February 2, 2009.

Cannon, Ellen, "The Near Credit Card Future," www.bankrate.com, March 12, 2009.

"CEO Forum: AmEx Chief Talks About Credit's Future," www.usatoday.com, April 8, 2008.

Chan, Andrew, "What the Credit Card Bill of Rights Means for Consumers," www.boston.com, May 21, 2009.

"Chase Paymentech and Revolution Money Expand RevolutionCard Merchant Acceptance," www.reuters.com, March 30, 2009.

Chu, Kathy, "Results Revive Old Questions, Add Others?" *USA Today*, Friday, May 8, 2009.

Clifford, Catherine, "Credit Card Delinquencies Hit Index Record," www.cnnmoney.com, March 10, 2009.

"Connecting People to Their Money," www.visa.com

Connors, Nick, "Consumer Payment Trends Favor Plastic Payment Methods," www.creditunions.com, October 27, 2008.

"Contactless Payment by Mobile Phone — Technically Possible, Work Continues on Requirements for Commercial Use," press release, www.creditsuisse.com, April 29, 2009.

Cooper, James C., "Consumer Spending Has Further to Fall," *BusinessWeek*, February 16, 2009, p. 14.

Cooper, James, "C., "Waiting for 'Credit Easing' to Kick In," *BusinessWeek*, February 9, 2009, p. 10. "Credit Card Giants Settle Antitrust Suit," www.upi.com, October 15, 2008.

"The Credit Cardholders' Bill of Rights: Balanced Reform," press release, www.maloney.house.gov, January 15, 2009.

"Credit Card Debt Statistics," www.money-zine.com

"Credit Card Industry Aims to Profit from Sterling Players," www.seekingalpha.com

Dash, Eric, "The Last Temptation of Plastic," www.nytimes.com, December 7, 2008.

Dash, Eric, "Credit Card Companies Willing to Deal Over Debt," www.nytimes.com, January 3, 2009.

Dash, Eric, "Consumers Feel the Next Crisis: It's Credit Cards," www.nytimes.com, October 29, 2008.

"The Debit Card Debate," www.fdic.gov, Spring 2006.

"Decline in Credit Card Spending Weighs in on Visa, MasterCard," www.money.cnn.com, January 16, 2009.

"Declining Credit Card Growth 2008: A (Leaking) Glass Half Full," www.insidearm.com, December 4, 2008.

deMause, Neil, "Congress Pushes for Credit Card Relief,"www.money.cnn.com, January 19, 2009.

Diaz, Sam, "Intuit Unveils Mobile Phone Credit Card Processing," blogs.zdnet.com, May 21, 2009.

DiscoverFinancial.com

"Discover vs. Visa and MasterCard," www.ecommerce-journal.com, September 24, 2008.

"Discover Card Goes Mobile," press release, www.discoverfinancial.com, January 15, 2009.

"Discover May Get $4 Billion from Visa, MasterCard Lawsuit," www.inside-arm.com, September 25, 2008.

"Discover Reaches $2.75 Billion Settlement with Visa, MasterCard," www.paymentsnews.com, October 27, 2008.

"Downgrading Visa, MasterCard, Heartland Payment Systems on Rapidly Worsening U.S. Bank Card Data – KeyBanc Capital Comments," www.rtt.news.com, January 16, 2009.

Dwinnell, Tate, "Visa vs. MasterCard: Comparing Credit Card Behemoths," www.seekingalpha.com, April 29, 2008.

"Economics of Participating in the Visa System," www.visa.com.

Ellis, David, "American Express Earnings Plunge 79%," www.money.cnn.com, January 26, 2009

Ellis, David, "Banks' Future Woes in One Word: Plastic,"http://www.money.cnn.com/, March 10, 2009.

"E-payments Boost Visa Earnings," *San Jose Mercury News*, February 5, 2009, p. 2C.

Evans, David S. and Schmaensee, Richard in *Paying With Plastic: The Digital Revolution in Buying and Borrowing*, MIT Press, 2005.

"Facts About Interchange," www.visa.com.

"Financial Cards In India," www.euromonitor.com, April 2008.

FirstData.com

Fitch, Stephanie, "Visa: The Future of Money," *Forbes Magazine*, October 27, 2008, p. 84-94.

Flaherty, Anne, "Big Changes in Store for U.S. Credit Cardholders," www.news.yahoo.com, May 20, 2009.

"Form 8-K for Discover Financial Services," www.biz.yahoo.com, October 28, 2008.

"Freedom Flat Tax Act," www.opencongress.org, February 12, 2009.

"Fueling Small Business Success," www.visa.com.

Fuscaldo, Donna, "Discover Launches Biodegradable Credit Card, Others to Follow," www.foxbusiness.com, January 15, 2009.

Google.com

Gale, William, "Business Taxes and the Flat Tax," www.brooksings.edu, March 7, 1996.

Gaffen, David, "Credit Card Trends Remain Negative," www.wsj.com, May 6, 2008.

Geer, David, "Cell Phone + Contactless Credit Card = Yak, Wave, Buy," www.creditcards.com, January 11, 2008.

Genuth, Iddo, "Future Credit Card," www.thefutureofthings.com, May 18, 2007.

Gerson, Emily Starbuck, "15 of the Coolest (and Weirdest) Credit Card Gadgets and Accessories," blogs.creditcards.com, January 8, 2008.

Glater, Jonathan, "Colleges Profit as Banks Market Credit Cards to Students," www.nytimes.com, January 1, 2009.

Goldberg, Daniel S., "E Tax: The Flat Tax as an Electronic Credit VAT," University of Maryland School of Law, www.papers.ssrn.com, September 5, 2005.

"Google Checkout vs. PayPal," www.bestshoppingcartreviews.com, April 17, 2009.

Graybow, Martha, "Discover Seeks $6Bln in Case vs Visa, MasterCard," June 9, 2008.

"An Overview," Bankcard Today, October 28, 2008.

Groos, Caleb, "Small Business Credit Card Rates Hiked and Limits Slashed; Prospects for Reform?" blogs.findlaw.com, May 12, 2009.

"The Growing Army of the Unemployed," *BusinessWeek*, February 16, 2009.

Hansel, Saul, "Visa Introduces a Credit Card in a Phone," www.nytimes.com, April 10, 2009.

Harvill, Erica, "MasterCard PayPass Adds Ease and Simplicity to More Transactions Worldwide Surpassing 50-Million Issued Milestone," www.mastercard.com, March 19, 2009.

"Higher Fees Could Be Rainmakers for the Bank Card Networks," www.digitaltransactions.net, March 17, 2009.

Holahan, Catherine, "Credit Cards Lose Ground to PayPal," www.articles.moneycentral.msn.com, December 12, 2007.

Holfich, Peter, "Retail Strategies Go Back to Basics," www.theasianbanker.com, January 14, 2009.

Holmes, Elizabeth, "April Store Sales Seed Recovery Hopes," *The Wall Street Journal*, May 8, 2009, p. B1.

"How Do Online Buyers Pay?" www.emarketer.com, January 14, 2009.

"How Federal Reserve Changes Affect Credit Card Industry," www.lowcards.com, December 18, 2008.

"History of Visa," www.visa.com.

"InformationWeek 500 Ranking Highlights Discover® MotivaSM Card Launch and Mobile Banking Demonstration," press release, www.discoverfinancial.com, October 1, 2008.

"H.R. 216: *Government Credit Card Abuse Prevention Act of 2009,*" www.govtrack.us.com, January 13, 2008.

"Increasing Trust in Payments," www.visa.com.

Jewell, Mark, "Visa Profit Rises 35 Pct Despite Recession," www.istockanalyst.com, February 4, 2009.

"KeyBanc Lowers Estimates and Price Targets on Payment Processors MasterCard (MA) and Visa (V)," www.streetinsider.com, January 15, 2009.

Kim, Ed, "Visa, MasterCard Risk Ramped Up Competition," www.seekingalpha.com, March 27, 2008.

Kon, Michael, "Lower Growth for Card Companies," www.morningstar.com, October 24, 2008.

Kuchinskas, Susan, "PayPal's Grown-Up Growth Outlook," www.internet-news, March 6, 2008.

Lagorio, Juan, "Visa, MasterCard Cut Costs Amid Slowing Revenue Growth," *International Business Times*, February 16, 2009.

Larsson, Naya, "MasterCard Worldwide Remains Top Corporate Card and Expense Services Provider According to *Global Finance Magazine*," www. mastercard.com, March 12, 2009.

Levitin, Adam, "The Credit Cardholders' Bill of Rights," www.creditslips. com, February 25, 2008.

Lieber, Ron, "American Express Kept a (Very) Watchful Eye on Charges," www.nytimes.com, January 31, 2009.

Lieber, Ron, "Credit Card Companies Go to War Against Losses," www. nytimes.com, January 31, 2009.

Lillis, Mike, "Is 2009 the Year of Credit Card Reform?" www.washington-independent.com, December 23, 2008.

Lynch, David, "U.S. Must Face Years of Sluggish Growth," *USA Today,* May 8, 2009.

MacDonald, Jay, "What Will Credit Cards Look Like in 25, 50, and 100 Years?" www.creditcards.com, February 17, 2009.

Mann, Ronald J., "Charging Ahead – The Growth and Regulation of Payment Card Markets," Cambridge University Press, 2006.

Mantell, Ru and Lillis, Mike, "Is 2009 the Year of Credit Card Reform?" www.washingtonindependent.com, December 23, 2008.

MasterCard.com

Martin, Patrick, "Your Android Phone Will Replace Your Wallet," www.an-droaidmobilephone.org, October 9, 2008.

"MasterCard Incorporated Reports Fourth-Quarter and Full-Year 2008 Financial Results," press release, www.mastercard.com, February 5, 2009.

"MasterCard Offers SMEs Tailored Payment Projects to Power Their Businesses," www.mastercard.com.

"MasterCard PayPass Adds Ease and Simplicity to More Transactions Worldwide Surpassing $50 Million Issued, www.mastercard.com

"MasterCard Settles Litigation with Discover Financial Services; Will Take a Third Quarter Net After-Tax Charge of $515.5 Million," PRNewswire, October 27, 2008.

McAfee, Seamus, "Heartland Data Breach Damages Still Mounting," www.creditcards.com, April 1, 2009.

McCool, Grant and Lagorio, Juan, "Discover, Visa and MasterCard Settle Antitrust Suit," www.reuters.com, October 24, 2008.

McCune, Erin, "eBay Analyst Day: PayPal World Domination," www.paymentsviews.com, March 11, 2009.

McDonald, Jay, "What Will Credit Cards Look Like in 25, 50 or 100 Years?" www.creditcards.com, February 17, 2009.

McFadden, Leslie, "5 Smart Credit-Card Moves in 2009," www.biz.yahoo.com, January 16, 2009.

McHenry, Justin, "Trends in Business Credit Cards for 2009," www.small-biztrends.com, January 17, 2009.

"Membership Rewards® Program from American Express Adds Practical Rewards for Tough Economic Times," press release, February 18, 2009.

"Merchant Solutions Enables Tens of Thousand of Asian Businesses to Accept Electronic Payments in First Year," www.firstdatacorp.com, November 19, 2008.

"Merchants Vent Their Ire Over Checkout Pricing on Google Site," www. digitaltransactions.net, March 25, 2009.

Metzger, Tyler, "Prepaid Cards Glitter in Gloomy Economy," www.credit-cards.com, December 17, 2008.

Miller, Peter, "Which Way Credit Trends in 2009?" www.debtcity.com, November 19, 2008.

Mitchell, Daniel J., Ph.D., "A Brief Guide to the Flat Tax," backgrounder, The Heritage Foundation, July 7, 2005.

Molina, Marcus, "MasterCard Reports Strong Growth Across the Latin American and Caribbean Region in 2008," February 11, 2009.

Murray, Jean, "Credit Card Legislation to Support Small Business Borrowers," www.about.com, September 23, 2008.

"New Bankruptcy Laws," www.lendingtree.com, August 6, 2007.

Newville, David, "The Future of Credit Card Regulation," www.newamerica. net, October 14, 2008.

Nazareno, Analisa, "Prepaid, reloadable payment cards for immigrants roll out," www.creditcards.com, September 19, 2008.

Nilson Report, The, April 2009.

"No Government Bailout on Personal Debt," www.lowcards.com, September 11, 2008.

"NSBA 2009 Small Business Credit Card Survey," National Small Business Association.

Palmer, Kimberly, "The End of Credit Card Consumerism," *U.S. News and World Report,* August 2008.

PayPal.com

"PayPal Merchant Service Major Source of Growth for eBay," www.seekingalpha.com, July 22, 2008.

"PayPal Growth Counterbalances eBay's Slower Growth," www.seekingalpha.com, July 24, 2009.

PCISecurityStandards.com

Plymouth, Kathi and Martin, Jodi, "Bill Payment Trends: Major Shifts in Consumer Behavior Require Comprehensive Planning, www.firstdata.com, 2009.

Prater, Connie, "Poll: Nearly 3 in 4 Feel Need More Credit Card Regulations," www.creditcards.com, August 18, 2008.

"Predictions for the Credit Card Industry 2009," www.lowcards.com, January 8, 2009.

"Report: Turkish Credit Card Market Expects Strong Growth," www.atmmarketplace.com, July 22, 2008.

Reuters, "Losses for Credit Card Companies Could Top $70 Billion," economictimes.indiatimes.com, January 3, 2009.

RevolutionMoney.com

Robertson, Jordan, "Credit Card Data Breach Could be Among Largest," *San Jose Mercury News,* January 22, 2009.

"Robust Growth in Malaysian Credit Card Market," www.btimes.com, November 5, 2008.

Rogak, Lisa, "Many Airlines Now Accepting Plastic for In-Flight Purchases," www.creditcards.com, August 4, 2008.

Rued, Marisol, "Internet Ready: American Express Wants Credit Card Growth Among Middle-Class Mexicans," www.thefreelibrary.com.

Ruysdael, Kai, "Visa's IPO Not Feeling the Crunch," www.marketplace. publicradio.org, February 25, 2008.

Saha-Bubna, Aparajita, "MasterCard's Profits Fall as Sales Drop," *The Wall Street Journal*, May 2-3, 2009, p. B3.

Sasseen, Jane, "The Bailout Is Broken," *BusinessWeek*, February 9, 2009, p. 21.

Scheeres, Julia, "When Cash Is Only Skin Deep," www.wired.com, November 25, 2003.

Schmith, Scott, "Credit Card Market: Economic Benefits and Industry Trends," U.S. Department of Commerce, International Trade Administration, March 2008.

Schulz, Matt, "Credit Cards Around the World: China," www.creditcards. com.

Scott, Mark, "Britain's New Rescue," *BusinessWeek*, February 9, 2009, p. 28.

"Shares of MasterCard (MA), Visa (V) Tumble on Analyst Comments Following Weak Card Data," www.streetinsider.com, January 16, 2009.

Simpson, Bill, "Visa's IPO Fueled by International Growth and U.S. Shift to Electronic Payments," www.seekingalpha.com, March 20, 2008.

Simon, Jeremy M., "Credit Card Balances Jump as Economic Woes Deepen," www.creditcards.com, March 6, 2009.

Simon, Jeremy, "Fed Report: Banks Continue to Tighten Lending Standards," www.creditcards.com, February 2, 2009.

Simon, Jeremy, "Fed Report Shows Plunge in Credit Card Use," www. creditcards.com, January 8, 2009.

Simon, Jeremy, "More Wacky, Wonderful Payment Card Innovations and Gadgets," blogs.creditcards.com, October 31, 2008.

Simon, Jeremy, "9 Predictions for Credit Cards in 2009," www.creditcards. com, December 30, 2008.

Neha Singh and Amiteshwar Singh, "Credit-card Industry May Cut $2 Trillion Lines: Analyst," www.reuters.com, December 1, 2008.

"Sizing the Market," Bankcard Today, October 28, 2008.

Snyder, Andrew, "Regulation Bonanza: Next Up, the Credit Card Industry," www.todaysfinancialnews.com, March 31, 2009.

Son, Hugh, "Visa's Net Rises 28% in First Report Since Record IPO (Update 3)," www.bloomberg.com, April 28, 2008.

"Sprint Processes GoPayment for SMBs," www.vpico.com, May 21, 2009.

Tan, Georgette, "MasterCard to Take a 52.5% Stake in Processing Company Strategic Payments Services," www.mastercard.com, February 18, 2009.

"Tightening Security," Bankcard Today, October 28, 2008.

"Tougher Rules for Credit Card Issuers in Works," www.i2credit.com, February 1, 2009.

Tozzi, John, "Credit Card Reform: Are You a Business, Consumer, or Both?" www.businessweek.com, May 1, 2009.

Tozzi, John, "House Passes Credit Card Reform, but Leaves Out Small Business Cards," www.businesweek.com, April 30, 2009.

"Trends in the U.S. Credit Card Industry," www.paymentsnews.com, April 1, 2008.

"Trends for Credit Card Possessors Discussed in Wall Street Transcript Report Business Services"; Education Issue, biz.yahoonews.com, February 11, 2009.

"Tuxedo Launches the UK's First Corporate MasterCard Prepaid Card," www.smartcardstrends.com, February 1, 2009.

"Understanding a Visa Transaction," www.visa.com.

"Urban Trust and First Data Partner to Offer Prepaid Cards," www.first-datacorp.com, March 3, 2009.

"U.S. Credit Card Losses Will Be Meaningfully Higher in 2Q," www.researchcap.com, May 5, 2009.

"U.S. Credit Card Industry Moving into Uncharted Territory," www.seekingalpha.com, May 21, 2008.

"U.S. House Approves Credit Card Reform Measure," www.washingtonpost.com, April 30, 2009.

"U.S. Regulators Adopted Sweeping New Credit-Card Rules," www.smartcardtrends.com, December 18, 2008.

Valentine, Kevin, "Senator Specter Introduces Flat Tax Act of 2009," www.wearecentralpa.com, March 31, 2009.

Vekshin, Alison, "Democrats Propose Measure to Limit Credit Card Abuses (Update 1)," www.bloomberg.com, January 15, 2009.

Verizon.com

Visa.com

Visa Corporate Report

"Visa to Launch Payments Processing Venture Targeting International High-Growth Geographies," press release, www.visa.com, December 2, 2008.

"Visa Profit Rises 35 Percent Despite Recession," www.istockanalyst.com, February 3, 2009.

"Visa Second Quarter Earnings Beat Estimates," www.news.moneycentral.msn.com, April 29, 2009.

"Visa Showcases the Future of Money at New York Event," www.visa.com, September 25, 2008.

Wade, Will, "The Future Shape of Payments Is Anything but Flat," *American Banker,* February 24, 2009.

Walsh, Mark, "Case Details Revolution Health's Growth and Future," www.mediapost.com, April 29, 2008.

Wattanajantra, Asavin, "RSA Europe: The Growth of the Underground Hacker 'Economy,' " www.itpro.co.uk, October 28, 2008.

Webb, Tim, "Cashless Society by 2012, Says Visa Chief," www.independent.co.uk, March 11, 2007.

Weil, Judy, "MasterCard on Brazil and Europe," www.seekingalpha.com, February 9, 2009.

Weston, Liz Pulliam, "7 Credit Card Trends that Can Cost You," www.moneycentral.msn.com, (undated).

Whitney, Meredith, "Credit Cards Are the Next Credit Crunch," www.wsj.com, March 11, 2009.

Wikipedia.org

Woolsey, Ben and Schulz, Matt, "Credit Card Statistics, Industry Facts, Debt Statistics," 2006-2009, www.creditcards.com, January 13, 2009.

Wolfe, David, "PayPal Responds to Rapid Evolution in Smart Phones," www.americanbanker.com, January 20, 2009.

Wong, Vince, "Google Checkout Becoming as Bad as PayPal," www.bitterwallet.com.

Yared, Georges, "KKR and First Data: Why a Payment Processor is Worth $26 Billion," www.bloggingstocks.com, April 1, 2007.

2009 Small Business Credit Card Survey, National Small Business Association, www.nsba.biz.

2009 U.S. Industry Outlook, "When Times Get Tough, the Tough Go Back to Basics," Nielsen, December 2008.

Lightning Source UK Ltd.
Milton Keynes UK
UKOW042104171012

200755UK00002B/152/P